INTRODUCING LIVING THINGS

Revised Nuffield
BIOLOGY
TEXT 1

Published for the Nuffield Foundation
by Longman Group Limited

Longman Group Limited
London
Associated companies, branches, and representatives
throughout the world

First published 1966
Revised edition 1974
Copyright © The Nuffield Foundation, 1966, 1974
ISBN 0 582 04601 7

Design and art direction by Ivan and Robin Dodd

Filmset in 11 on 12 point Century Schoolbook
by Keyspools Ltd., Golborne, Lancashire
and made and printed in Great Britain
by Butler and Tanner Ltd., Frome and London

REVISED
Nuffield Biology

Text 1
INTRODUCING LIVING THINGS

General editor
Grace Monger

Editors of this volume
Grace Monger
Margaret E. Tilstone

Contributors
C. D. Bingham
D. M. Dallas
W. H. Dowdeswell
B. A. C. Dudley
John A. Hicks
Alison Leadley Brown
Grace Monger
Margaret E. Tilstone

Cover photograph by
Peter Jackson/Bruce Coleman.

Contents

Foreword

It is ten years since the Nuffield Foundation undertook to sponsor curriculum development in science. The subsequent projects can now be seen in retrospect as forerunners in a decade unparallelled for interest in teaching and learning not only in but far beyond the sciences. Their success is not to be measured simply by sales but by their undoubted influence and stimulus to discussion among teachers—both convinced and not-so-convinced. The examinations accompanying the schemes of study which have been developed with the ready cooperation of School Certificate Examination Boards have provoked change and have enabled teachers to realize more fully their objectives in the classroom and laboratory. But curriculum development must itself be continuously renewed if it is to encourage innovation and not be guilty of the very sins it sets out to avoid. The opportunities for local curriculum study have seldom been greater and the creation of Schools Council and Teachers' Centres have done much to contribute to discussion and participation of teachers in this work. It is these discussions which have enabled the Nuffield Foundation to take note of changing views, correct or change emphasis in the curriculum in science, and pay attention to current attitudes to school organization. As always, we have leaned on many, particularly those in the Association for Science Education who, through their writings, conversations, and contributions in other varied ways, have brought to our attention the needs of the practising teacher and the pupil in schools.

This new edition of the Nuffield Biology *Texts* and *Teachers' guides* draws heavily on the work of the editors and authors of the first edition, to whom an immense debt is owed. The first edition, published in 1966, was edited by Professor W. H. Dowdeswell, organizer of the Biology project which carried out the trials in schools of the original draft materials. The authors of the first edition were:

Alison Leadley Brown Texts I and II and Teachers' Guides I and II

C. D. Bingham Texts I and II and Teachers' Guides I and II

A.K. Thomas Text III and Teachers' Guide III

A. Ellis Texts III and IV and Teachers' Guides III and IV

A. Darlington Texts III and IV and Teachers' Guides III and IV

P. J. Kelly Text V and Teachers' Guide V

The new edition contains a preponderant part of these authors' material, either in its original form or in edited versions. They are credited among the authors of the new edition but their wider contribution in providing a firm basis for further developments must be gratefully acknowledged here.

I particularly wish to record our gratitude to Grace Monger the General Editor of this new series. It has been her responsibility to organize and coordinate this revision and it is largely through her efforts that we have been able to ensure the fullest cooperation between teachers and the authors.

As always I should like to acknowledge the work of William Anderson, our publications manager, and his colleagues, and, of course, to thank our publisher, the Longman Group, for continued assistance in the preparation and publication of these books. I must also record our debt to those members of Penguin Education who were actively involved in the preparation of the books until a late stage in their production. The contribution of editors and publishers to the work of the course team is not only most valued but central to effective curriculum development.

K. W. Keohane
Co-ordinator of the Nuffield Foundation Science Teaching Project

Preface
to the first edition

To the pupil

You have probably been told that biology is the study of living things. This is true but, by itself, the statement does not tell you what biology is really about. In this course we want to do more than just teach you how living things function; we want you to understand why scientists wish to know about life and how they set about finding out biological truths.

Scientists must be curious; they must be prepared to form tentative guessing answers to the questions they ask themselves; and they must be able to test these guesses. We have tried to guide you through this process, to show you why you should be curious, what kinds of questions you should investigate, and how you should devise and carry out experiments. Experiments are not intended to prove things you already know; they are to investigate whether something does or does not happen so that you can form hypotheses which, themselves, can be tested by further experiments. Thus, a negative result may be as important as a positive one. We have also tried to indicate how you should use the results you get; how you should test them further and how you should relate them to the questions you posed yourselves.

By the end of this course we hope you will know not only more about living things, particularly man, but also more about how to study living things both in the laboratory and in their natural state.

To the teacher

The essence of this Nuffield course in biology is 'science for all'. In devising it, our intention has been to provide a balanced and up-to-date view of the subject suitable for pupils who will leave school at the age of sixteen and do no more formal biology. For some it will also provide a jumping-off point for further study at A-level.

The course has been built around a number of fundamental themes. Such issues as the relationship of structure and function, adaptation, and the interaction of organism and environment are discussed again and again in different contexts throughout the five-year period. The course is designed to foster a critical approach to the subject with an emphasis on experimentation and enquiry rather than on the mere assimilation of facts. In terms of a conventional syllabus this means that less factual matter is included. This in itself is no bad thing, provided the principles of the new teaching are accepted and the methods used are in sympathy with the aims of the course. In order to foster this outlook, a *Teachers' guide* has been produced. This is closely cross-referenced to the *Text* and contains copious notes on teaching methods, also much technical matter relating to practical work, including additional experiments. In short, our aim has been to produce not only a new syllabus, but, more important, a new approach to teaching.

The course falls clearly into two parts; the first two years which can be regarded as introductory, and the remaining three which constitute the next (intermediate) phase. The introductory phase is characterized by a broad general approach to the subject. In the intermediate phase the treatment becomes more quantitative with greater emphasis on experimentation and reasoning.

Preface
to the second edition

The most important feature of the second edition of the Nuffield Biology *Texts* is the part played in their revision by teachers and pupils who have had the experience of following the course to O-level. Before any decisions were made to change the materials first published in 1966, a long process of evaluating how far they had succeeded in fulfilling the aims and intentions of the original Project took place. From this exhaustive investigation into the use of the first edition, the editors and authors of this new edition have drawn invaluable help in deciding how to present the material and to take into account the criticisms and suggestions of practising teachers. In this sense, the second edition can be seen as the result of a further stage of the trials on which the original materials were based.

The introductory phase (two years) is now in one volume instead of two but the intermediate phase is still covered in three volumes. The subject matter has undergone considerable rearrangement and re-editing. Where topics have been developed at greater length in some cases, or have been shortened in others, these changes are based on the reports and requests from schools which were received during the evaluation described above.

The intentions of the course, however, remain the same and the aim of the revision has been to refine the materials and to bring them up to date where this is necessary.

Grace Monger, General Editor

Investigating living things

Most of us are interested in the things we see and find out-of-doors and would like to know more about them for our own enjoyment. Many of us would like to feel that we could add something to what is known about an animal or plant in which we have been interested, or about living things in one of our favourite places. But we might think that there was nothing we could say that had not been said already, nothing we could observe that had not been observed, recorded, and written about a thousand times before.

We would be quite wrong to think this. For as we learn how to look at things and how to use books, we begin to see that there are a great many gaps in man's knowledge. Did you realize, for instance, that we know practically nothing about the life of a whale? You might say that whales, because they live in the sea and often very deep down, are difficult animals to observe, and you would be right, but it would be equally true to say that we know very little about some of the commonest animals which are easy to observe but which have, so far, escaped the more careful attention they may well deserve. Some starlings leave us to fly to warmer countries in autumn, while others at the same time arrive in this country to spend the winter with us. Still others stay here all the year round. What is the reason? Is it always the same starlings that migrate or stay at home? Biologists have not yet answered these questions.

You can investigate some living things and find out where and how they live and how they behave. Although you will be dealing with animals in this chapter, do not forget that plants are living things and that investigating them is as interesting and worth while as investigating animals.

You may wonder where you can find some animals to investigate. Obviously this will be easier for some than for others but the following list will give you ideas of the places where you might look. *Figure 1* shows one of these places.

1 Under stones in a rockery or garden or under any pile of rubble that has been about for some time.
2 Under any piece of wood or sacking that is lying about.
3 In any damp shaded place where moss or a few other plants have managed to grow.
4 In a gutter which has become choked with moss and debris.
5 Under any rubbish that has been lying about for some time.
6 Under the bark of logs or felled trees.
7 In the leaf litter under trees.
8 In the crevices in an old brick or dry stone wall (be careful not to damage the wall).
9 In the school pond or any other nearby pond.
10 On the shore in rock pools, under the seaweed growing on rocks, or in the seaweed washed up on the tide line.
11 In the soil on any piece of land you can dig over.

Figure 1
A habitat which may be in your school grounds.
Photograph, John May.

When you start to look for some animals, try to disturb the habitat (the place where the animals live) as little as possible. It is very easy to destroy a habitat and cause the death of the animals and plants living there. Make sure that you replace stones, logs, etc. as close as you can to their original position and collect only a few animals of each type. When you have examined the animals in the laboratory, make sure that you return them to their habitat. If you follow these simple rules you will keep the habitat for the plants and animals living there and for the enjoyment of others who investigate it after you.
If necessary, you will be given more detailed instructions on how to collect the animals from the habitat.

Before you take your animals back to the laboratory to examine them, make a record of as many things about the habitat as you can. Some of the things you could record are:

1 Is it dry or damp? Will it normally be in this state?
2 How much sun does it get during the day?
3 Is it exposed to all weathers or does it get some shelter?
4 Is it disturbed regularly or are you the first person to examine it for some time?

1.2 What have you found?

When you take your animals back to the laboratory, one of the things that you may like to do is to identify them. In order to do this you will have to look at them very carefully and you may need to use a hand lens.

1.21 Using a hand lens

The kind of hand lens usually used in a biology laboratory magnifies an object either eight or ten times and since you will often be using hand lenses it is important that you should learn how to do this properly. You can practise by looking at the skin on the back of your hand.

1 Hold the lens about 8 cm from one eye, keeping the other eye open if you can.
2 Now, holding your other hand palm side down, bring it up to the lens until it is about 16 cm away.
3 Continue to move it slowly. As you do so, the skin on the back of your hand will gradually become clear (be in focus). You will notice that not only is your skin much magnified but you will be able to see things on the surface of the skin, such as hairs and wrinkles, which you could not see before.

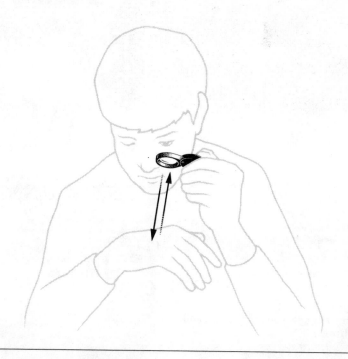

Figure 2
The right way to use a hand lens. Note that you should bring the thing you are looking at nearer to the lens and not the other way round.

With the help of your hand lens look carefully at each of your animals. Then, using your own ideas, put the animals with similar features into the same group. To make closer identification you will need to use books that you have in the laboratory or can get from the library. The pictures in *figure 3* are to help you to identify some of the smaller animals. Make a list of the animals that you identify.

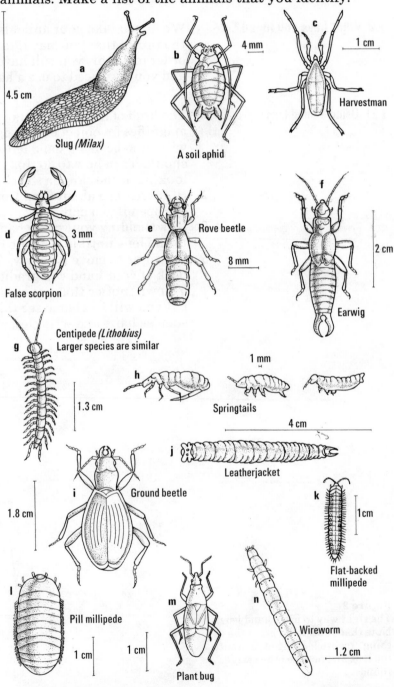

4.5 cm

a

Slug *(Milax)*

b

4 mm

A soil aphid

c

1 cm

Harvestman

d

3 mm

False scorpion

e

Rove beetle

8 mm

f

2 cm

Earwig

g

Centipede *(Lithobius)*
Larger species are similar

1.3 cm

h

1 mm

Springtails

4 cm

j

Leatherjacket

i

Ground beetle

1.8 cm

k

1 cm

Flat-backed millipede

l

Pill millipede

1 cm

m

1 cm

Plant bug

n

Wireworm

1.2 cm

Figure 3
Some animals you may find in soil or litter.

Introducing living things

Figure 3 (continued)

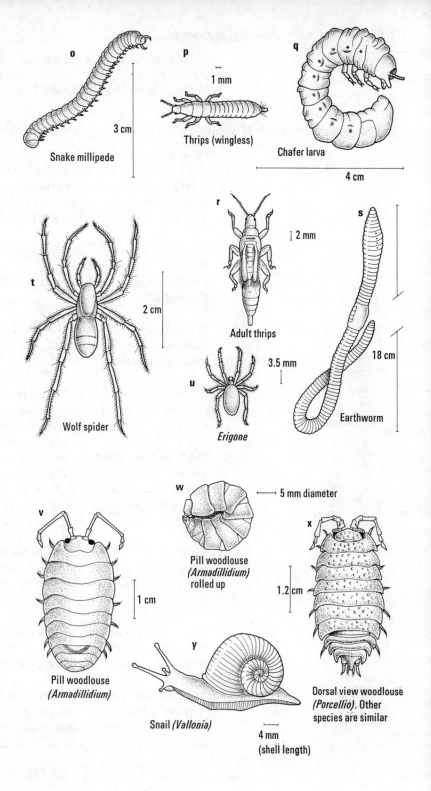

o
3 cm
Snake millipede

p
1 mm
Thrips (wingless)

q
4 cm
Chafer larva

r
2 mm
Adult thrips

s
18 cm
Earthworm

t
2 cm
Wolf spider

u
3.5 mm
Erigone

v
1 cm
Pill woodlouse
(Armadillidium)

w
5 mm diameter
Pill woodlouse
(Armadillidium)
rolled up

x
1.2 cm
Dorsal view woodlouse
(Porcellio). Other
species are similar

y
Snail *(Vallonia)*
4 mm
(shell length)

Now that you have identified some of the animals you may like to take one or two and look more closely at how they are built or how they behave. Return the rest to their habitat.

1.31 Looking at the structure of animals

In order to look at the structure of the animals you will need to use your eyes and a hand lens very carefully. Do not think that because you have seen a picture and been able to identify it from this that there is nothing else to be seen. The following questions will help you to look carefully. *at an animal you have found*

Q1 What shape is it? Can you suggest any reason why its shape may fit it to the place in which it lives? For example, the long, thin shape of the earthworm helps it to burrow through the soil.

Q2 Is the body in one piece or is it divided into two or more main sections? For example, the body of a spider is divided into two sections.

Q3 Is the body divided into small, more or less identical, segments (rings)? For example, the body of a maggot is in one piece but this is divided into segments.

Q4 What is the outer surface like? It may be hard or soft, dry or moist, smooth or hairy. Do you think that this may be an advantage to the animal or affect the way in which it lives? For example, the shell of a tortoise gives it protection and the moist skin of a frog means it has to live in damp places.

Q5 What colour is the animal? Do you think that this may be of any use to it? For example, the colour of a greenfly (an aphid) helps to camouflage it whilst it is feeding on plants.

Q6 Has the animal any legs? How many? Are all the structures which look like legs actually used for movement? If not, can you see how they are used?

Q7 Are there any other structures attached to the body? Can you see where they are attached and how they are used? For example, the fly has a pair of wings attached to the middle section of its body and they are used for flying.

Q8 Has the animal any eyes? How many are there and where are they found?

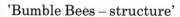

Q9 Can you see the animal's mouth? Is there anything around it or is it protected in any way?

Q10 Can you see any other features?

When you have answered these questions, use the information to write a description of the animal, adding any sketches which will make this clearer.

The following is the beginning of a long, detailed description by a biologist. It is an example of a good, clear piece of writing about an animal.

Figure 4
The queen of the buff-tailed bumble bee.

'Bumble Bees – structure'
'The queen of the Buff-tailed Bumble Bee is a large insect, whose body may be 1 inch [2.5 cm] long and the wings $1\frac{3}{4}$ inches [3.75 cm] across. The body is thickly covered with hair, which for the most part is black; but at the front of the thorax is a band of deep yellow, and there is another belt of yellow across the front of the abdomen. The tail of the abdomen is yellowish, but not so yellow as the other bands, and the hairs of the tail usually have white or cream tips. In old specimens the tail may look almost white.'
From Alan Dale, Patterns of life.

1.32 Finding out how they move

While you have been writing the description of your animal you will probably have seen it moving. Indeed, it may have been moving about so much that writing your description was difficult. Three sets of observations follow. These may not be of the animal which you have been examining but they illustrate three types of movement. When you have looked at these you will be able to use them as a guide when examining other animals.

Finding out how an earthworm moves
1   Put a large earthworm on to a sheet of brown paper and notice which is the 'head' end of the worm and which is the 'tail' end. Find the saddle (clitellum) nearer the head end and the red blood vessel which shows through the skin and runs all along the back.
2   Watch how the worm moves forward by extending the front (anterior) end of its body. As it does so, this part of the worm becomes much longer and thinner. At the same time note what is happening to the rest of the body.
3   Listen carefully as the worm moves over the paper.
4   Gently run your fingers along the sides of the worm from head to tail and then from tail to head. How do you think the anterior end of the worm grips whilst the rest of the body is drawing up?

5 Put the worm on a damp, glazed tile. How does it move now? Can you explain the reason for the difference?

Finding out how a slug moves
1 Take a clean polythene bag, moisten the inside, and then put some slugs into it.
2 Lay the bag flat on the bench and wait for the slugs to start moving. When the slug moves along does its body change shape like that of the earthworm?
3 Carefully pick up the bag and lift it so that you can see the underside of the slug. What can you see? Is this in any way similar to the movements of the earthworms?

Finding out how an insect moves
Most insects have two ways of moving but we are only going to look at how an insect walks. (What is the other way in which many can move?) You can use either a stick insect or an aphid (greenfly or blackfly). If you look at an aphid you will have to use your lens very carefully.
1 Put the insect on a piece of paper. Watch the insect moving. Which of the following things happens?
a The front legs both move at the same time and in the same direction.
b The middle legs both move at the same time and in the same direction.
c The hind legs both move at the same time and in the same direction.
2 Now look carefully at the three legs on one side. Which of the following things happens when the insect moves?
a All three legs move at the same time.
b The front and middle legs move at the same time.
c The front and back legs move at the same time.
d The middle and back legs move at the same time.
e The legs all move at different times.
3 Now look at the legs on the other side. Do they move in the same way?
4 Look at all the legs. Can you see which legs move at the same time?
5 Make a simple model of an insect, using Plasticine and pins as shown in *figure 5*. Try removing some of the pins. What is the smallest number needed to keep the model upright? Can you relate this to what you have learnt about the way in which the insect moves?

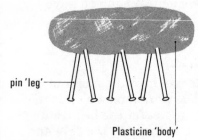

pin 'leg'

Plasticine 'body'

Figure 5
A simple model of an insect.

1.33 Looking for evidence of their activities

Sometimes it is possible to say that an animal lives in a particular habitat without actually seeing the animal. This is because many of them leave evidence of their activities which are easily recognizable. *Figures 6 to 11* show some examples of this evidence. Can you name the animal responsible in each case?

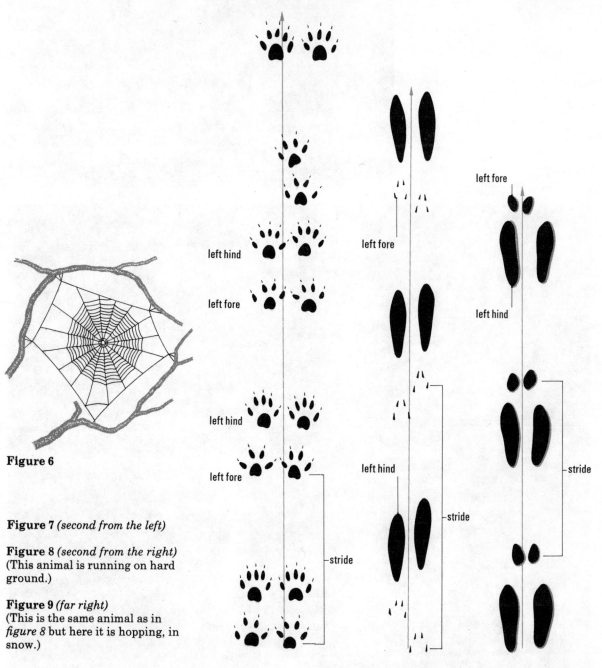

Figure 6

Figure 7 *(second from the left)*

Figure 8 *(second from the right)*
(This animal is running on hard ground.)

Figure 9 *(far right)*
(This is the same animal as in *figure 8* but here it is hopping, in snow.)

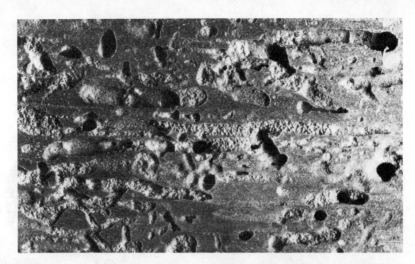

Figure 10
Photograph, Rentokil Ltd.

Figure 11
Photograph, R. J. Corbin.

You will be able to collect other examples of evidence of the activities of animals.

You will have realized by now that any type of animal lives only in certain places and you may have wondered why this is so. For many animals the answer is very complicated but this should not prevent you from making some suggestions or hypotheses which may explain your observations. You can then perform some experiments which will test whether your hypotheses are correct.

Let us take an example. Many of you will have found woodlice when you first searched for animals. These are very common and are found under leaves, logs, or stones, or in similar places.

Q1 What features do these places have in common?

Q2 Can you therefore make a hypothesis about the places where woodlice live?

Test your hypothesis by setting up a tray like the one shown in *figure 12*. Release a number of woodlice in the tray and then leave it for 24 hours. At the end of this time, find out where the woodlice have gone.

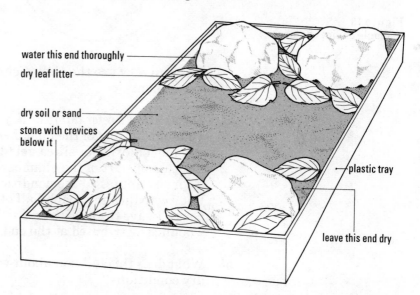

water this end thoroughly

dry leaf litter

dry soil or sand

stone with crevices below it

plastic tray

leave this end dry

Figure 12
Tray for testing what conditions woodlice prefer.

Depending on where the woodlice have collected you will be able to say whether these results suggest that your hypothesis is true, but there could have been other things which influenced this result. For example, they may have been attracted to the damp leaves as a source of food. To be really sure that the hypothesis is correct you need to investigate one condition at a time. Can you devise ways of setting up the tank in order to do this?

Another way of testing one condition at a time is to use a special piece of apparatus called a choice chamber. There are several types of choice chamber but a simple one is shown in *figure 13*.

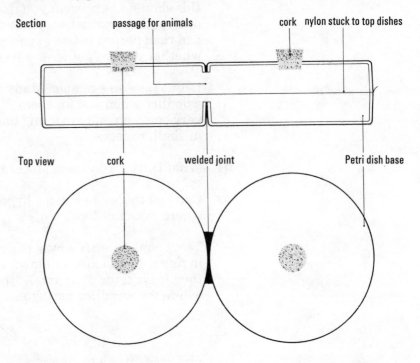

Figure 13
A simple choice chamber.

You can now use the choice chamber to carry out some experiments.

Do woodlice prefer damp or dry conditions?

1 Take the bottom of the chamber. Put water in one half and calcium chloride or silica gel (drying agents) in the other.
2 Put the top part of the chamber into the bottom and leave it for 10 minutes so the conditions can become steady.
3 Put 5 woodlice into each half of the top and replace the corks. Leave the apparatus for 20 to 30 minutes. How are the woodlice distributed at the end of this time?

Q3 What does this tell you about whether they prefer damp or dry conditions?

Do woodlice prefer dark or light conditions?

1 Put equal amounts of water into the bottom sections of the choice chamber.
2 Put the top of the chamber into position and put 5 woodlice into each half.
3 Cover one half with a piece of black paper and leave the apparatus for 20 minutes. How are the woodlice distributed at the end of this time?

Q4 What does this tell you about whether they prefer light or dark conditions?

You should now be able to say whether your original hypothesis is correct.

Although woodlice have been suggested in these experiments there is no reason why you should not use other animals and design experiments of your own, using the choice chamber or a modification of it.

Q5 How could you find out whether millipedes prefer warm or cold conditions?

Q6 How would you find out whether the preference of woodlice for darkness is stronger or weaker than their preference for dampness?

How Darwin investigated the earthworm

On a cold December day in 1831, His Majesty's Ship *Beagle* sailed from Plymouth on a voyage around the world which was to last five years. On board was a young naturalist, Charles Darwin, then only 22 years old. Darwin, whose portrait as an older man you see in *figure 14*, was a born collector. In those days it was fashionable to collect things and many of our most famous museum collections were made then. But there was a difference in the purpose of the collections which Darwin made on his long voyage. To him the animals, plants, and fossils meant more than mere specimens in a museum. After he returned from the voyage he was to spend many years in studying and writing about the things which he had found. He wrote many books which were to make him, even during his lifetime, one of the most famous scientists that have ever lived. His most famous book was *On the origin of species by means of natural selection,* but he wrote many others which were the result of many years of patient study, often of quite ordinary animals and plants such as barnacles, orchids, and even earthworms.

No one would imagine that such a small and comparatively simple animal as an earthworm could be of any importance. Yet Darwin spent many years in collecting information about worms, and finally he wrote the famous book called *The formation of vegetable mould through the action of worms.* The book, which may seem to us to have rather a dull title, was destined to become a best-seller.

Figure 14
Charles Darwin.
Photograph reproduced by kind permission of the President and Council of the Royal College of Surgeons of England.

Although Darwin had the advantage of living in a comfortable country house in Kent with a large garden in which he could work undisturbed, most of his observations and records about earthworms could have been made by anyone with the necessary patience, time, and interest.

How did Darwin set about his work and what is there that we can learn from his methods? First of all we find that quite small things caught his attention. He noticed that if fallen leaves in a wood were swept away, the whole surface of the ground was strewn with worm casts and that the fine earth of the casts (the result of the worms actually eating the soil and passing it through their bodies) eventually buried the leaves falling once more on top of them. This layer of fine soil and leaves Darwin called vegetable mould.

In his book Darwin tells of a narrow path running across the lawn in his garden, which was paved with flat stones. For several years the path was weeded and swept, but the worms threw up many casts and more weeds grew. In the end, the gardener gave up weeding the path and thirty years later, a layer of turf over three inches (about 8 cm) thick had formed on top of the paving stones.

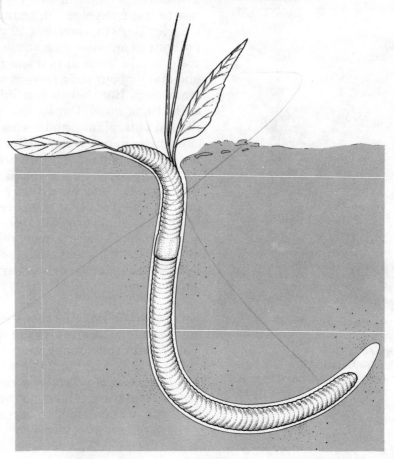

Figure 15
A common earthworm drawing a leaf into its burrow (the burrow is shown in section).

In the same way Darwin noticed that stones in fields became buried in the course of a few years, by a layer of soil brought up to the surface by worms. Although he kept careful measurements of the depth of vegetable mould overlying the stones, and checked and re-checked his observations, he wanted to find out more about the way in which the worms worked. So he kept some worms in large jars of soil in his study so that he could watch more carefully the way they made their burrows, plugged the entrances, and made their soil casts.

From his various records Darwin found out that the layer of vegetable mould deposited on the surface of the ground by worms could be as much as an inch (2.5 cm) thick in a matter of a few years. As a means of checking his results, he resolved to collect, for a certain number of days, all the worm casts within a square yard (about a square metre) of ground. He dried the soil of the casts then weighed it. In this way he arrived, by careful experiment, at the astonishing conclusion that in the course of ten years worms could cover a field with soil to a depth of nearly two and a half inches (just over 6 cm).

It is easy to understand, therefore, how even quite large stones can become covered with soil cast up by worms. In the same way, worms can be responsible, in the course of hundreds of years, for completely burying ancient buildings. In fact, Darwin showed that earthworms have played a most valuable part in preserving such buildings, even the beautiful mosaic pavements, cooking pots, and coins left by the Romans nearly two thousand years ago. By covering them with soil the worms have protected the valuable historical remains from the action of frost and rain.

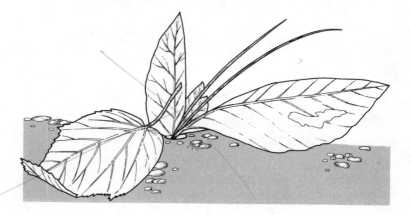

Figure 16
The burrow plug of an earthworm. Leaves and other material are drawn into the burrow's entrance.

Many gardeners nowadays make use of earthworms to help them cultivate the ground so that they do not have to dig it themselves. By covering the soil with a layer of decaying leaves and other vegetable matter, the gardener can leave the rest to the worms. They will gradually draw all the vegetable matter down into the ground, and at the same time, their worm casts will cover the mould, forming rich soil in which plants can grow. Thus the worms will do all the hard work!

Another way in which worms help the gardener and the farmer is by breaking up the soil into smaller particles. Have you ever noticed how fine the soil of a worm cast is? Try rubbing a cast between your fingers. The burrows which worms make are also useful in that they help the roots of plants to grow by allowing air and rain water to enter the soil. The fine soil produced on the surface in their casts helps to cover up the seeds of plants and provides them, too, with the conditions they need for growth. To quote the words of Darwin himself, 'The plough is one of the most ancient and most valuable of man's inventions; but long before he existed the land was in fact regularly ploughed, and still continues to be thus ploughed by earthworms.'

Why measure?

2.1 Biology is a science

Biology is a scientific subject and the people who study it are scientists. Scientists are interested in finding the answers to questions by experiment. They do experinents, using them to try to answer a question. The sort of questions they ask themselves are:
How does a bee find its way to and from the hive?
Can a bull really see the colour red?
How do plants grow?
What causes cancer?

Scientists set up suitable experimemts and watch what happens. Then they describe both the experiment and the results so that others will know about them, and finally they decide whether or not the results allow them to answer the original question.

Althcugh doing an experiment is important, nothing can be discovered unless the scientists watch what happens. This might seem easy, but it is not. For instance, there have been three spelling mistakes so far in this chapter. Did you notice them?

Now try answering these two questions without looking around the room:
How many ceiling lights are there in the room?
What is the colour of the door handle?

It is easy not to notice something and this can happen when you are watching experiments just as easily as it can happen at any other time.

2.2 Misleading cases

Even when you are watching carefully, sometimes you can be mistaken about what you think you see. Find out how easily this can happen with the following tests.

Q1 Are the lines between the V shapes in *figure 17* of equal length?

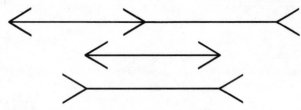

Figure 17

Q2 Are all the lines in *figure 18* straight?

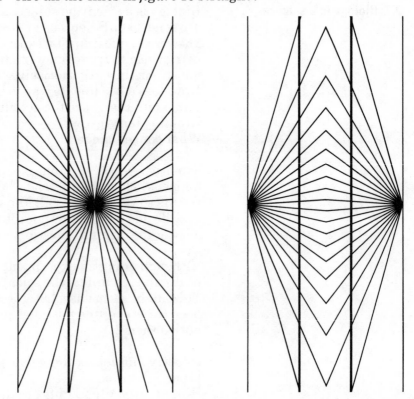

Figure 18

Q3 Are the dotted circles in *figure 19* the same size as each other?

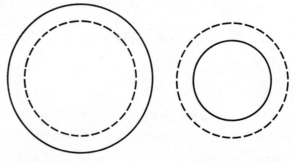

Figure 19

Introducing living things

Q4 In *figure 20* are the circles in the centre the same size as each other?

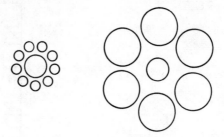

Figure 20

Q5 Are the shapes in *figure 21* the same size as each other?

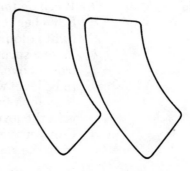

Figure 21

Q6 Is the inset shape in *figure 22a* a square?
Is the one in *figure 22b* a circle?

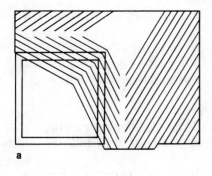

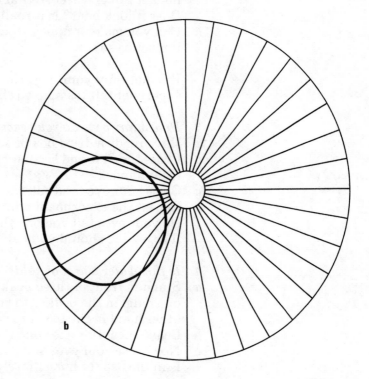

a

b

Figure 22

Q7   Are the diagonal lines in *figure 23* parallel?

Figure 23

2.3 Things to do

1 *When do the bottoms of the beakers appear to be level?*
a Fill a beaker with water and place it on a white or light-coloured surface.
b Take a second similar but empty beaker and, looking down from above, raise it until the bottom of the empty one seems to be level with the bottom of the beaker full of water.
c Now look from the side and see how much difference there is in the levels of the two beakers.
d Can you arrange the beakers so that the bottoms look level from the side as well as from above?

2 *Is the ruler really bent?*
a Hold a wooden ruler at an angle and put it into a sink or bucket full of water so that only half of it is submerged. Does it look bent? Is it really bent?
b Can you make it appear to be straight even when you know it must be so?

It is not only your eyes that can deceive you. Try this next experiment. It tests how reliable your sense of touch is.

3 *How many peas do you seem to feel?*
a Have ready a dried pea or something else that is round.
b Cross the first and second fingers of either your left or your right hand.
c Close your eyes and roll the pea between the crossed fingers.
d Repeat the experiment while looking at the one dried pea. Do your eyes tell you one thing, and your fingers another? If so, which should you believe, and why?

4 *Is this water warm or cold?*
a Stand in front of three beakers of water. Place a finger from one hand in the beaker to your left and a finger from the other hand in the basin to your right.
b Leave them there for one minute.
c Now close your eyes and put both fingers in the water which is in the beaker immediately in front of you.

Next time you drink a cup of tea that feels just nicely warm when you taste it, dip a clean finger into it and see whether or not it feels the same temperature. Then answer the question: 'Is the tea hot or is it warm?'

5 *How warm is the stream of air?*
Place one dry and one wet hand in the stream of air coming from a hair dryer or fan heater. Is the stream of air warm or is the stream of air cold? Where is the air warmest – near the dryer or away from it?

2.4 Misleading cases solved

How can the questions asked in section 2.2 and those that follow be answered accurately?

It seems that the only way is to take measurements and not to be influenced by appearances. This is one reason why scientists measure.

Turn back to the diagrams in section 2.2.

Take the measurements indicated below and then answer again the questions that are asked for each of the diagrams.

Figure	Measurement
17	Measure the lengths between the V shapes
18	Devise your own way of testing these lines
19	Measure the diameter of the dotted circles
20	Measure the diameter of the circles in the centre
21	Measure some suitable lengths of these shapes
22a	Measure the angles of the inset figure
22b	Put a coin inside the inset figure
23	Devise your own measurements for testing these diagonal lines.

Next, repeat experiments 4 and 5 (section 2.3), this time using a thermometer to measure the temperature. Are the answers different from those you decided upon when you were using only your hands? What is the best answer to each of the questions asked in these two experiments?

2.41 Common measurements

So far you have measured lengths, angles, and temperatures. These are called *physical quantities*.

In order to answer the second part of question 6 (page 19), you used a standard shape (a coin) to test for a circle. On each of the other occasions you used a *standard unit* of measurement. *Table 1* lists the physical quantities you have measured so far and their standard units. It also includes a quantity, mass, which you have not measured yet.

Physical quantity	Standard unit	Symbol
length	a metre (parts of a metre	m
	are centimetres and	cm
	millimetres)	mm
temperature	a kelvin	K
mass	kilogramme (parts of a	kg
	kilogramme are grammes)	g
angles	degree of a circle	°, e.g. 90°

Table 1

The position is a little complicated. Scientists in most of the countries of the world have agreed to use the same standard units. These are called SI units, which stands for *Système International d'Unités*. The first three standard units given in the table are also SI units.

But other units *can* be used and a number of alternatives exist. For example, temperature can be measured in degrees Fahrenheit (°F) or degrees Celsius (°C) as well as in kelvins (K). In fact, you will measure it in degrees Celsius. Similarly, the standard unit for the measurement of length can be the inch or it can be the metre.

On the other hand, the table gives 'degree of a circle' as the standard unit for measuring angles. This is not an SI unit, although everyone agrees about the basis it uses – the division of a circle into 360 parts or degrees.

The table does not give all the standard units, of course, but only those which you are likely to find useful.

Note: A measurement that is often used for giving the size of very small things is μ (the Greek letter mu). This means 10^{-6} m. In this *Text*, it is used for the first time to give the actual size of the egg shown in *figure 55* (page 53).

2.5 Measurement at work

Another reason why scientists measure is that many questions they are asked are concerned with quantities. Here is a sample of such questions. See if you can establish the answers.

1 *How much?*
a The amount of food a mouse eats each day is equal to its own mass. How much food has to be provided each day for the mice present in the classroom?

b How much food can a hungry locust eat at one time? Make a note of the time taken for the locust to eat its fill. How long would it take for 20 hungry locusts to eat a whole lettuce?

c Two piles of potatoes have been set out in the classroom. Suppose each pile consists of all the potatoes grown by one plant. Which plant produced the greater number of potatoes? Which plant produced the greater amount of potato?

d How much water is there in food? Devise some experiments of your own but talk to your teacher about your ideas before starting any of them.

2 *How warm?*

Some animals live amongst the leaves of a tree, some amongst grass, some in soil, and some in water. Which of these places is the warmest today?

Try to repeat the test at a different time of day and see if you get the same result.

Which animals experience the greatest range of temperature at this time of year, and which the least?

3 *How many?*

a How many strides do you have to take to walk 10 metres? If one stride takes one second how many metres would you walk in one minute? How many minutes would it take you to walk 1000 metres?

b Some leaves of named plants have been provided for you. How many times longer than broad is each type of leaf? Is the answer very different for large and for small leaves of the same plant?

c How many centimetres does a woodlouse walk in 20 seconds? Does it walk faster or slower than a fly maggot? If a woodlouse and a maggot had a walking race, how many centimetres apart would they be after 1 minute and how many centimetres would the slower one have walked?

4 *How long?*

a For how long can you hold your breath?

b How long is your forearm?

c How long are the leaves on the shoot that were brought into the classroom today?

d What is the circumference of your cranium?

e How long does a cat live?

2.6 Getting organized

For some questions there is not one answer, but many. This is true, for instance, of each question under the heading 'How long?' and it is only fair to make a note of all the different answers.

2.61 Organizing your own results

Lengths of leaves

1 Make a list of the lengths of all the leaves on the shoot (section 2.5, experiment *4* part *c*).
2 List the increasing lengths, starting at the top with the smallest measurement.
3 Make a note of the length of *a* the smallest leaf, and *b* the largest leaf. This is the *range* of length of leaves.
4 What is the length of a leaf that is exactly in the middle of this range? This is the *median* length of the leaves on the shoot.
5 Add all the lengths and divide the final total by the number of leaves measured. This is the *mean* length of the leaves on the shoot.
   The *range* and the *mean* are often used to describe variable results such as you have for these leaves. The *median* is not so useful.
6 Organize the data into a frequency table by copying and completing *table 2*. Under the second column, score each leaf against the class it belongs to by making a mark.

Length of leaf	Score of leaves in each class	Number of leaves in each class
0 to 2.0 cm		
2.1 to 4.0 cm		
4.1 to 6.0 cm		
6.1 to 8.0 cm		
8.1 to 10.0 cm		
10.1 to 12.0 cm		
12.1 to 14.0 cm		
14.1 to 16.0 cm		

Table 2
Frequency table for length of leaves on the shoot of laurel.

7 Draw on graph paper a histogram of your results. This is another way in which results can be organized.

Cranium measurements

1   Measure the circumference of your own cranium.
2   Organize the class results into:
a   A list in order of increasing measurements, with the smallest measurement at the top.
b   A frequency table – decide with your teacher the most suitable groupings for your measurements.
3   Draw on graph paper a histogram of the results in the frequency table.
4   Calculate and write on the graph paper *a* the range, *b* the median value, and *c* the mean value for the circumferences of the craniums in your class.
5   How does the circumference of your own cranium compare with:
a   the mean value in your class?
b   the median value in your class?

2.62 Organizing the results of others

Number of flowers on bluebell heads
Parts *A* and *B* of *table 3* represent the number of flowers on the flower stalks of bluebells on two different sites. Is there any noticeable difference in the number of flowers per head amongst the bluebells found at the two sites? Tackle the answer to the problem by means of the steps set out below.

A Site: sunny bank

7	12	9	5	9	7	10	6
7	8	3	7	11	4	15	8
11	9	7	6	9	5	6	18
8	13	4	8	7	10	15	2
9	7	8	10	13	6	9	10

B Site: shady, moist woodland

3	6	11	7	4	3	6	2
5	4	2	4	3	5	5	3
8	4	3	1	6	2	4	5
5	9	4	5	4	7	3	6
10	3	8	4	7	3	5	4

Table 3
Number of flowers in the heads of bluebells.

a Work in pairs, one taking the results of *table 3A* and the other of *table 3B*. Pool your results at some suitable stage.
b Draw up a frequency table of results.
c Draw on graph paper a histogram of the results in *table 3A* and *table 3B*.
d What is the range, the mean, and the median for bluebell heads taken from these two sites?
e What is the most frequent number of flowers in the heads of bluebells on the two sites? This most important frequent number is known as the *mode*.
f What conclusions can you now draw from the results? Using the *range*, *mean*, *median*, and the *mode*, describe the differences. Notice that any differences that might be present do not stand out clearly until after the above steps have been taken. This is the purpose behind organizing the data.

Mass of domestic hen
Draw a graph of the results shown in *table 4*. They are an average of a bird's development, taken from a study of a number of hens.

Table 4
Growth of a domestic fowl.

Age from hatching (days)	1	30	50	75	100	125	175	225	300
Mass (g)		50	300	575	1150	1900	2300	2900	3000	3000

Age on reaching maturity: 160 days

Use this information and the graph to answer these questions:

Q1 What was the mass of the fowl when it first reached maturity?

Q2 By how many grammes did the mass of the fowl increase between the time it reached maturity and the time it reached the final adult mass?

Q3 How many days passed between the time the fowl reached maturity and the time it reached the final adult mass?

Q4 About two-thirds of the mass of the fowl is lost when the bird is 'dressed' ready for cooking. In that case, what is the mass of the 'oven-ready' bird, killed on first reaching maturity?

Q5 On this basis what must have been the live mass of a fowl whose dressed ('oven-ready') mass is
a 1350 g (about 3 lb)
b 1800 g (4 lb)?

2.7 New measurements for old

Some years ago it was decided that all scientists should use the same units of measurement and, as a result, many measurements that are familiar will disappear, at least from scientific work. Such units as the 'mile', 'foot', 'pound', and 'ounce' will no longer be used. At present it is not at once clear if a horse, which can travel at 40 miles per hour, has a greater speed than a kangaroo which is capable of speeds up to 48 kilometres per hour.

If the tail of one mouse is $3\frac{1}{2}$ inches in length, is it longer or shorter than one 9.2 cm in length?

What is needed is a simple way of converting measurements made in one system of units (inches, for instance) into equivalent measurements in another system of units (metres). There are a number of ways in which this can be done. Two are given in *figures 24* and *25*, each concerned with converting degrees Fahrenheit into degrees Celsius, but these methods can be used to convert any equivalent measurements.

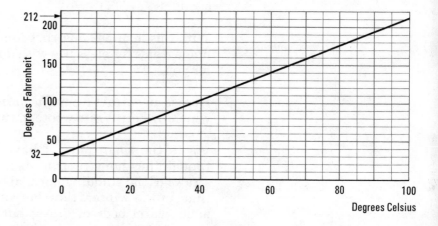

Figure 24
A conversion graph.

Use *figure 24* to answer the following questions.
a   The body temperature of humans is 98.4°F. What is the corresponding temperature in °C?
b   Comfortable room temperature is 20 °C. What is this temperature on the Fahrenheit scale?

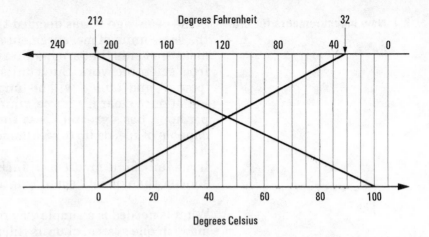

Figure 25
A nomogram.

Use *figure 25*, a *nomogram*, to answer the same two questions asked above. Compare the results obtained from the graph and from the nomogram. Which method do you prefer? The most accurate result is obtained when the graph and the nomogram are drawn large and with care, using a sharp pencil so that all the lines are fine. They must also be straight.

With this in mind construct some conversion graphs and nomograms for class use with data which will be provided for you.

Here is an extract from *The natural history of Selborne* which Gilbert White wrote in the eighteenth century.

'As to the small mice, I have farther to remark, that though they hang their nests for breeding up amidst the straws of the standing corn, above the ground; yet I find that, in the winter, they burrow deep in the earth; and make warm beds of grass; but their grand rendezvous seems to be in corn ricks, into which they are carried at harvest. A neighbour housed an oat rick lately, under the thatch of which were assembled near a hundred, most of which were taken; and some I saw. I measured them, and found that, from nose to tail, they were just two inches and a quarter, and their tails just two inches long. Two of them, in scale, weighed down just one copper halfpenny, which is about the third of an ounce avoirdupois, so that I suppose they are the smallest quadrupeds in this island. A full grown *Mus medius domesticus* weighs, I find, one ounce lumping weight, which is more than six times as much as the mouse above; and measures from nose to rump four inches and a quarter, and the same in its tail. We have had a very severe frost and deep snow this month. My thermometer was one day fourteen degrees and a half below the freezing point, within doors. The tender evergreens were injured pretty much. It was very providential that the air was still, and the ground well covered with snow, else vegetation in general must have suffered prodigiously. There is reason to believe that some days were more severe than any since the year 1739–40.'

(Letter XIII)

This is an example of a natural history classic in which, of course, the data are not given in SI units. You may meet other examples where information would have to be converted to SI units for comparisons to be made with more recent work. See, for example, page 7.

2.8 How big is 'big'?

Sometimes a word is used that makes measurement difficult. The word 'size' is one. How do you measure the size of a person, an earthworm, a potato, or a drawing?

Perhaps we would measure the height of a person, the length of an earthworm, the mass (or maybe the volume) of a potato, and the dimensions or even the area of a drawing. On each occasion we should be measuring some aspect of size, but as we should be taking a different measurement in each case this could lead to confusion, especially if we wished to compare the sizes of these things. Just as it has been necessary to standardize the units of measurement, so it is desirable to use a standard way of comparing sizes.

There is no satisfactory standard way of comparing the sizes of things which are different in shape. But for things which are similar in shape, and especially for drawings and photographs, there is a satisfactory standard method which is to use the *scale factor*. This is written as, for example, ' $\times 2$' or ' $\times \frac{1}{2}$'), alongside the drawing or photograph. ' $\times 2$' means that every linear measurement of the drawing or photograph is twice the length of the real object. When the scale factor is ' $\times \frac{1}{2}$' every linear measurement of the drawing or photograph is half the length of the real object.

Look through this book at the drawings which show a scale factor. From this and the length or height of the thing drawn, calculate its length or height in real life. Try this for three plants and for three animals.

Cells as units of life

3.1 Looking at the structure of living things

When a biologist wishes to study a new plant or animal, one of the questions he is likely to ask is 'What is its structure?'. This can be answered in various ways. The biologist may want to know about the chemicals of which the living thing is composed. He may be looking for the larger units which make up the body of the organism in the way that bricks make up a house when they are put together. Many different kinds of buildings can be constructed from the same kind of bricks, so if we study the great variety of living things in detail, we may find that they have some sort of building units in common.

Examine a piece of wood, first of all with the naked eye, and then with a hand lens.

Q1   What becomes visible when you examine the cut end with the hand lens?

Now study *figure 27* which shows a thin section cut across the tree trunk illustrated in *figure 26*, seen through a microscope.

Q2   What is the value of a microscope?

3.2 Using a microscope

A compound microscope makes it possible to magnify small objects or parts of tissues and see them more clearly. It usually has a range of lenses. By combining these in different ways you can choose the required magnification.

A good school microscope may be able to magnify up to about × 900, although you may often see all you need at a much lower magnification. The Background reading gives information about how even higher magnifications, up to × 200 000, are obtained.

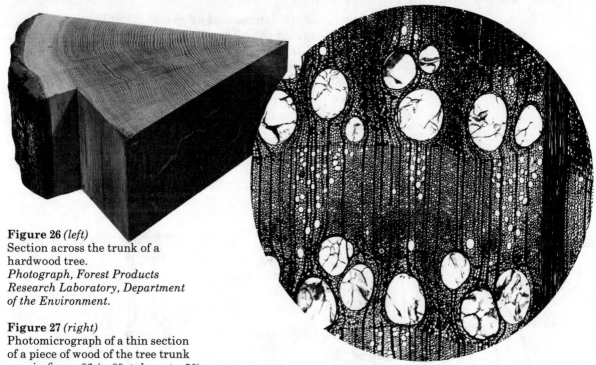

Figure 26 *(left)*
Section across the trunk of a
hardwood tree.
*Photograph, Forest Products
Research Laboratory, Department
of the Environment.*

Figure 27 *(right)*
Photomicrograph of a thin section
of a piece of wood of the tree trunk
seen in *figure 26*. (× 30; taken at × 50)
*Photograph, Forest Products
Research Laboratory, Department
of the Environment.*

3.21 The parts of a microscope

Details follow of the structure of a microscope and methods
of obtaining low and high power magnification, although
the high power may only be used in certain cases.

Microscopes vary in design but all rely on one principle for
producing a magnified picture or image. In some ways this
principle is similar to the one you may already have used in
working with a hand lens. A compound microscope is
designed to carry out two of these magnifications, one after
the other. In other words there is a first lens, the objective,
to produce a magnified image, and then a second lens, the
eyepiece, to magnify this image still further.

For many years the usual arrangement for lighting and
focusing was to have a platform (stage) fixed to the body of
the instrument and a tube to hold the lenses which moved
up and down for focusing. More recently, microscopes have
been produced with the tube fixed in one position and a
stage which moves up and down instead.

The essential parts of a modern microscope are as follows:
a A metal *body* which forms a base to which the other parts
are attached. Always lift a microscope by grasping the body
of the instrument.
b A *stage* to hold the slide with a central hole to let the light
through.
c A *light* below the stage or a *mirror* to reflect light.

d A *tube* to hold the lenses and keep out all light except that which comes from the object being viewed.

e *Objective:* the lens at the lower end of the tube. Different magnifications can be obtained by changing objective lenses and many microscopes have a rotating turret or nosepiece to hold two or more different objectives.

f *Eyepiece:* the lens in the top of the tube through which you see the image.

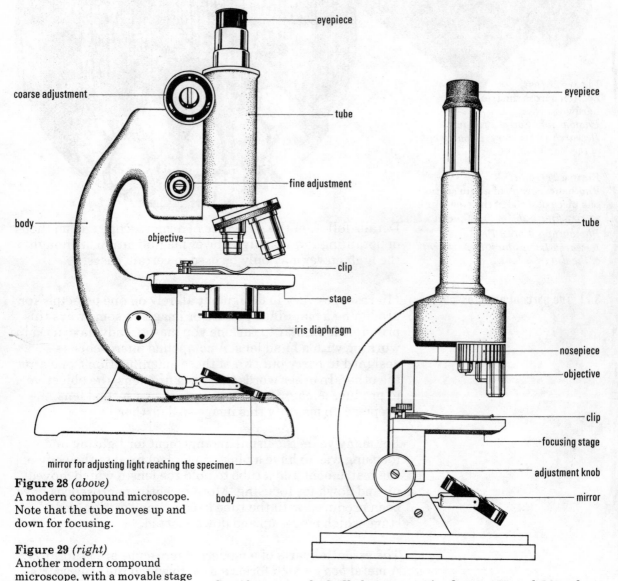

Figure 28 *(above)*
A modern compound microscope. Note that the tube moves up and down for focusing.

Figure 29 *(right)*
Another modern compound microscope, with a movable stage for focusing.

See if you can find all these parts in *figures 28* and *29* and on your own microscope as well.

3.22 Looking through a microscope

When you use a microscope you must follow carefully the instructions given here to get good results.

Using the low power magnification

1 If there is a mirror, arrange it so as to reflect the light towards the stage. Light may be from a bench lamp or from the sky (but not direct sunlight). Other microscopes may have a lamp built in below the stage.
2 Select the objective with the lowest power, that is, the one with the shortest mount and the widest lens.
3 Place your slide on the stage with the specimen you wish to examine near the centre of the hole. Hold it in position by the clips.
4 Look at the side of the microscope and turn the coarse focusing knob to bring the objective as close to the slide as you can without the two touching. (Remember that in some microscopes the focusing knob moves the tube, and in others the stage.)
5 Look through the eyepiece and turn the focusing knob to move the lens away from the slide. NEVER FOCUS BY MOVING THE LENS TOWARDS THE SLIDE WHILE LOOKING THROUGH THE EYEPIECE.

If you do move the lens and slide towards each other and miss the focus point, you are likely to crush the slide and lens when they meet. This can be an expensive mistake. But if you miss the focus while you are moving the slide and lens away from each other, you will merely reach a point at which the tube will travel no further. Do not try to focus the knob beyond this point. Start again by bringing objective and slide close together and then move them apart a little more slowly and look more carefully. You should eventually reach a point where you see a clear picture through the eyepiece. Try to keep both eyes open while you look down the instrument.

Using the high power magnification

1 Using low power, arrange the slide so that the specimen you wish to examine is in focus in the middle of the field.
2 Swivel the nosepiece round so that the narrower objective lens, on the longer mount, is in line with the microscopic tube.
3 Look down the eyepiece and, using the fine focusing knob if your microscope has one, wind the objective and slide away from each other very slowly. The specimen should come into view as you do this.
4 Ask for help if you have difficulty.

3.23 Making a temporary slide

Although it is possible to buy ready-made permanent microscope preparations, it is very easy to mount fresh specimens for temporary use; these are called temporary slides.

When you use a hand lens or a binocular microscope, you will find that light falls onto the specimen from the sides and above. Light is usually passed *through* the specimens viewed with a compound microscope.

At first it is easiest to study very small or transparent specimens, for example, yeast cells suspended in water, young moss leaves, or small plants and animals from ponds or aquaria. It is also useful to be able to study thin skins. Later, other methods of obtaining thin specimens will be described.

You are going to start by making a temporary slide of your cheek cells and then use your microscope, as described, to look at them. Look at the questions 1 to 7 at the end of this section and answer them as you go along.

Examining cells from inside the cheek
1   Gently scrape the inside of your cheek with a clean spatula.
2   Touch the surface of a clean microscope slide with the spatula so that a drop of saliva is transferred to the slide with the cheek scrapings. Add a drop of water.
3   Add a coverslip as described in *figure 30*.

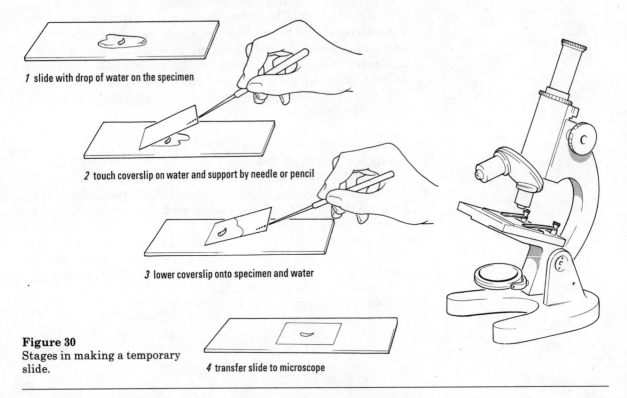

1 slide with drop of water on the specimen

2 touch coverslip on water and support by needle or pencil

3 lower coverslip onto specimen and water

4 transfer slide to microscope

Figure 30
Stages in making a temporary slide.

(Air bubbles may appear in your temporary mount, as circles with hard, black edges. Try to recognize them for the future. If there are only a few, examine another part of the slide but if there are many, start again.)

4 Repeat the mounting described above. Now add a drop of methylene blue solution instead of water.

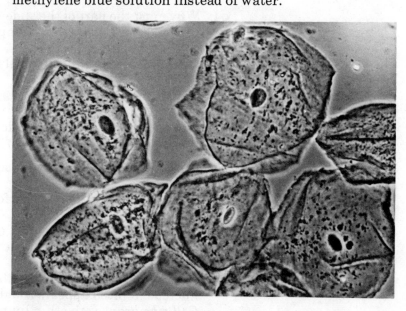

Figure 31
Cheek cells seen under the low power of a microscope (× 410). *Photograph, Harris Biological Supplies Ltd.*

Examining the 'skin' on leaves

1 Break a leaf sharply.
2 Pull apart the two halves. You can now pull off a layer one cell thick. Put this on a microscope slide.
3 Add a drop of water or dilute iodine solution and a coverslip as described in *figure 30*.

When you have examined some unstained and stained slides under the microscope, decide how the stains alter the appearance of the specimen.

Move one of the temporary mounts you have made from left to right on the stage of the microscope.

Q1 What do you notice about the movement of the image as you look at it down your microscope?

If the magnification of the eyepiece is marked × 4 and the objective is × 10, the microscope will have a total magnification of × 40. Now look at your microscope.

Q2 What magnification are you using
a with the low power objective in position?
b with the high power objective?

Alter the amount of light passing through the specimen. Ask how to do this if you are not sure.

Q3 Is it true that the more light there is the clearer the picture you will get?

Q4 Why must the part of the slide you wish to examine be in the centre of the field as you change from low to high power?

Q5 Does greater magnification always give a clearer picture?

Q6 Is staining useful?

Q7 If you have used two stains, do they act in the same way?

3.3 The parts of a cell

Here are some of the words used to describe the parts of a cell.

Protoplasm is a general term for all the living material in a cell.

Nucleus is the special region, often near the centre of the cell, which usually appears denser. This is the controlling region of the cell.

Cytoplasm is the name given to the protoplasm outside the nucleus. It often appears rough or granular.

Cell membrane is the thin living membrane enclosing the cytoplasm.

Vacuoles are cavities which may be present in the cytoplasm, especially in plant cells. They are usually filled with fluid.

Cell walls of dead, rigid material surround the cells of plants.

Chloroplasts are small, green structures containing chlorophyll present in many plant cells.

3.31 Comparing different kinds of cells

Using the method outlined on page 34, mount some cheek cells, onion cells, and young moss leaves and stain them so that the details of the cells are as clear as possible. (The onion cells are obtained by peeling off some of the outer skin covering the fleshy scale leaves inside the onion.)

Make clear, labelled drawings of what you can see in each case.

Copy *table 5* into your notebook, making it as large as possible, and complete it.

	Cheek cells	Onion cells	Moss leaf cells
What is the shape of the cell?			
What is the position of the nucleus?			
Are vacuoles present?			
If so, how many?			
Are chloroplasts present?			
If so, where in the cells?			
Is there a cell wall?			

Table 5
Comparing some different kinds of cell.

3.4 Other methods of making cells clearer

So far, you have only looked at thin specimens. You are now going to see what thicker specimens, such as the mustard seedlings shown in *figure 32*, look like under a microscope.

Figure 32
A mustard seedling in a clay pot, showing root tips at the correct stage for making a squash preparation.
Photograph, C. D. Bingham.

1 Cut off the end of a root tip provided and mount it in water as before.
2 Look at it under the low power of your microscope.

Q1 How much detail of the structure of the root can you see?

Because cells are tightly packed together in many living things, thinner preparations must be made if we are to see the arrangement of individual cells and their contents. Only in this way can enough light travel through the specimen.

Squashes
To view a root tip of onion, mustard, or cress carry out the instructions given in *figure 33*.

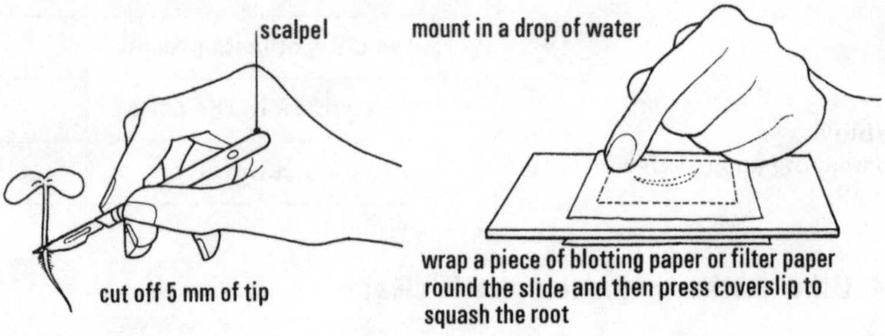

scalpel

mount in a drop of water

cut off 5 mm of tip

wrap a piece of blotting paper or filter paper round the slide and then press coverslip to squash the root

Figure 33
Making a preparation of mustard root.

Examine the root tip under the low power of your microscope.

Q2   Are all the cells the same shape?

Q3   Where are the largest cells?

Q4   Where are the smallest cells?

Q5   Do all the cells appear the same shade of grey?

Note that cells will look darker if they let through less light.

Q6   What ways can you suggest of making the cell contents more easily visible?

Sections
You may be given the opportunity to cut sections yourself or to examine ready-made sections.

When biologists want to cut many very thin sections they use a special instrument called a microtome, a little like a bacon slicer, and they often embed delicate specimens in a wax to prevent them from being broken. Sections are stained in a similar way to suspensions and skins.

Permanent slides
Biologists often buy or make permanent microscope slides of specimens.

To make these, it is necessary both to stain the specimens

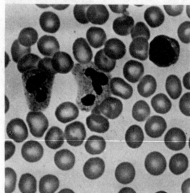

Figure 34
Photomicrograph of a permanent preparation of human blood.
Photograph, Harris Biological Supplies Ltd.

very carefully, and to remove water. They are put into special mounting materials which harden and remain clear for many years. Such preparations are used, for example, in the photomicrographs of blood (*figure 34*), of cork cells (*figure 42*, page 44), of a pollen grain (*figure 46*, page 46), and of bacteria (*figures 149c* and *160*, pages 151 and 166).

Q7   What advantages do these permanent slides have?

Examine the photographs in *figure 35* and answer questions 8 to 13.

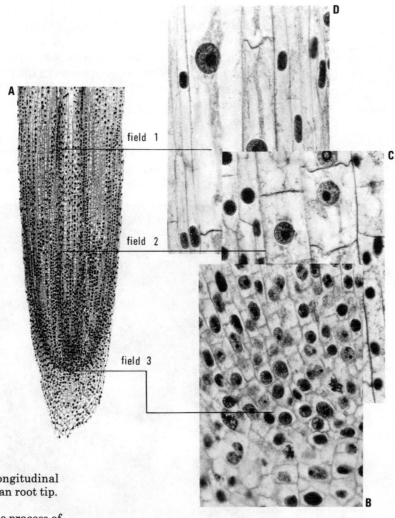

Figure 35
Photographs of a longitudinal section of broad bean root tip.
A (×25).
B Young cells in the process of dividing (×115).
C Cells growing larger (×115).
D Cells which have grown even larger by the development of a vacuole (×115)
Photographs, Harris Biological Supplies Ltd.

Q8   What do the cells appear to be doing in B?

Q9   In what way do the cells in C differ from those in B?

Q10 In D, the cells have altered even more. How?

Q11 Which cells are the youngest?

Q12 Which cells are the oldest?

Q13 At the tip there is a cap of slimy cells. What might their function be?

Microscopes then and now

Lenses for magnification have been made for over two thousand years, and by about A.D. 1400, simple spectacles were being used. The microscope was invented in Europe in about 1600 and many varied designs were introduced. At about the same time the telescope was also invented and this and some early microscopes were quite similar in design, using different lenses at each end. One early microscope was almost 2 metres long and 2.5 cm in diameter!

Many of the early users of microscopes were enthusiastic amateurs, eager to glimpse this world of the very small which became visible for the first time.

Figure 36
Portrait of Antonie van Leeuwenhoek. He is holding one of his microscopes.
By courtesy of the Wellcome Trustees.

Introducing living things

Figure 37
A facsimile of van Leeuwenhoek's original microscope in the University of Utrecht.
Photograph by courtesy of the Wellcome Trustees.

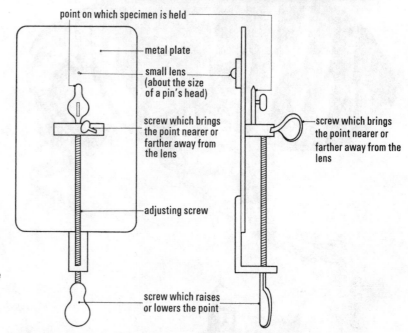

point on which specimen is held

metal plate

small lens
(about the size
of a pin's head)

screw which brings
the point nearer or
farther away from
the lens

screw which brings
the point nearer or
farther away from the
lens

adjusting screw

screw which raises
or lowers the point

Figure 38
Diagrams to show parts of Antonie van Leeuwenhoek's simple microscope. *(Left)* Front view. *(Right)* Side view.

One of these early experimenters was a Dutchman called Antonie van Leeuwenhoek (1632 to 1723) whose portrait you see in *figure 36*. He had the idea of mounting a high-powered lens (only 2 mm across) in a metal plate, instead of holding it in his hand. The object to be examined was mounted on the pointed rod, which could be moved up or down, and nearer or further away from the lens, by means of two screws. This arrangement overcame the problem of movement of the object. *Figures 37* and *38* show the structure of Leeuwenhoek's *simple* microscope.

Leeuwenhoek's hobby was making lenses and during his lifetime he constructed more than 240 of these rather crude little microscopes. The amount of magnification he obtained was about × 240, which is a great deal better than anything we can achieve with a hand lens today.

By modern standards, these early microscopes were poor. The image they produced was not very clear and the

magnification was quite low. An ordinary school
microscope will magnify objects 400 times without
difficulty, while good instruments have magnifications of
× 1000, or even more. But in spite of these difficulties,
Leeuwenhoek was able to see in such things as soil or
souring milk a whole new world of tiny living creatures
which no one before him had dreamed of. He called his
minute creatures 'little animals'. Today we call them
microbes.

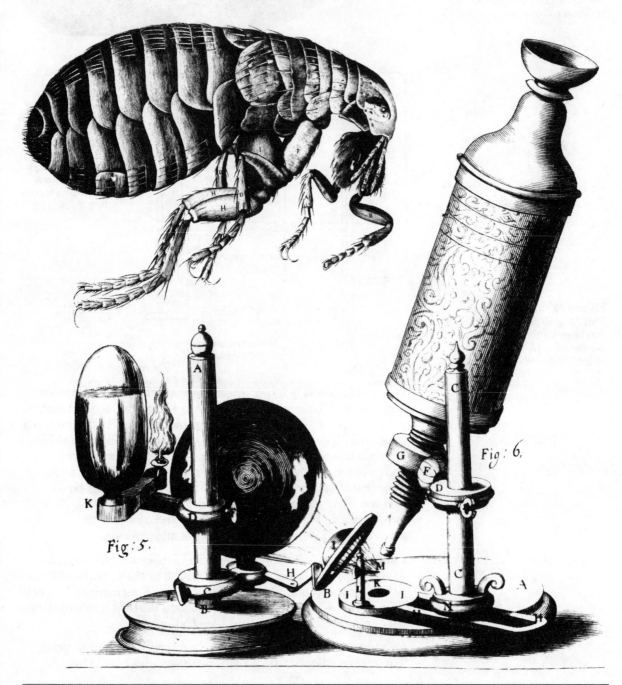

Fig: 5.

Fig: 6.

The kind of compound microscope we have today was used by Robert Hooke (1635 to 1703). He described his microscope in *Micrographia*, which includes many illustrations (such as *figure 39*) of objects which he had seen with the aid of this instrument (*figure 40*). In one chapter we find the following account of his examination of a piece of cork.

'I took a good clear piece of cork, and with a Penknife sharpen'd as keen as a Razor, I cut a piece of it off, and thereby left the surface of it exceeding smooth, then examining it very diligently with a *Microscope* methought I could perceive it to appear a little porous; but I could not so plainly distinguish them, as to be sure that they were pores . . . I with the same sharp Pen-knife, cut off from the former smooth surface an exceeding thin piece of it, and placing it on a black object Plate, because it was itself a white body, and casting the light on it with a deep *plano-convex Glass*, I could exceedingly plainly perceive it to be all perforated and porous, much like a Honey-comb, but that the pores of it were not regular . . . these pores, or *cells*, were not very deep, but consisted of a great many little Boxes . . . Nor is this kind of texture peculiar to Cork onely; for upon examination with my *Microscope*, I have found that the pith of an Elder, or almost any other Tree, the inner pulp or pith of the Cany hollow stalks of several other Vegetables: as of Fennel, Carrets, . . . Teasels, Fearn . . .&c. have much such a kind of Schema-tisme, as I have lately shewn that of Cork.'

His drawing of cork cells is reproduced in *figure 41* and you can compare it with a photomicrograph taken through a modern microscope (*figure 42*).

Hooke not only described and drew these 'cells', he also measured them:

'. . . I . . . found that there were usually about threescore of these small Cells placed end-ways in the eighteenth part of an Inch in length, whence I concluded there must be neer eleven hundred of them . . . in the length of an Inch, and therefore in a square Inch above a Million, or 1166400, and in a Cubick Inch, above twelve hundred Millions, or 1259712000 a thing almost incredible. . . .'

The Royal Society had received its Royal Charter in 1662 and Hooke became its first 'curator of instruments'. A famous botanist, Nehemiah Grew (1641 to 1712), was also a Fellow of the Royal Society and he used Hooke's microscope to make drawings accurate enough to be included in a modern textbook. See *figure 43*.

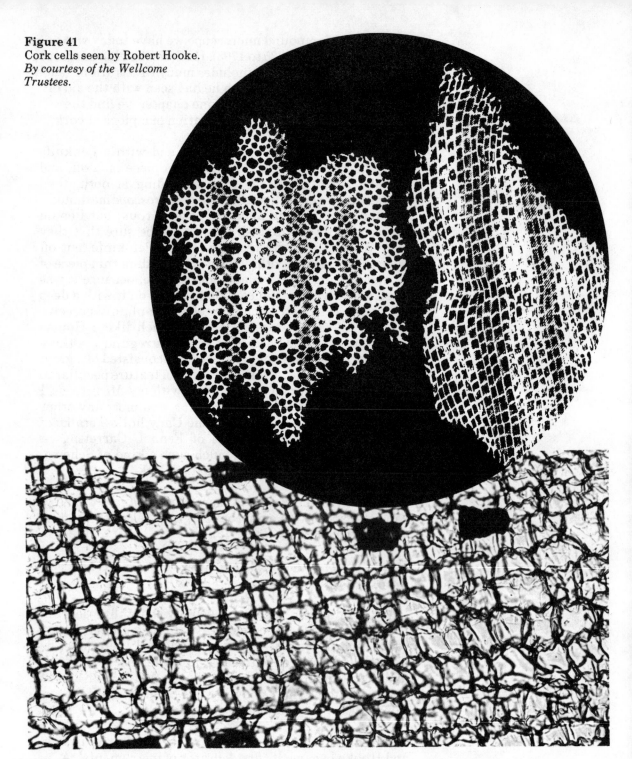

Figure 41
Cork cells seen by Robert Hooke.
By courtesy of the Wellcome Trustees.

Figure 42
A photomicrograph of cork cells taken through a modern microscope.

During the eighteenth and nineteenth centuries, the compound microscope was improved until, in basic structure, it resembled the microscope we use today, although our microscopes lack the beautiful decoration of many of the older designs.

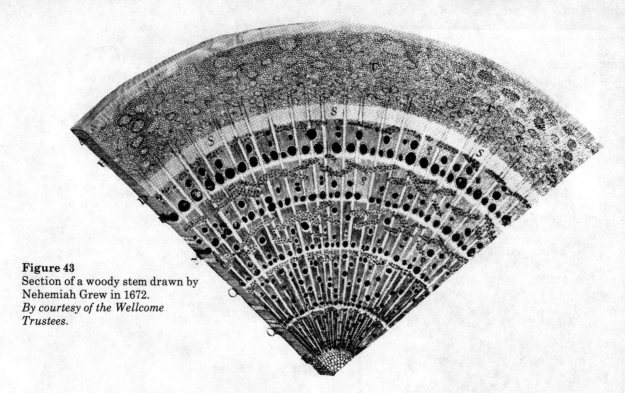

Figure 43
Section of a woody stem drawn by
Nehemiah Grew in 1672.
By courtesy of the Wellcome
Trustees.

Staining and lighting methods were also improved, but
however good these are, and however much a light
microscope is modified, it cannot normally magnify more
than × 1500, and × 400 is more usual.

In the late 1930s another kind of microscope was invented in
Germany.

Instead of using a beam of light it was found that a beam of
electrons could produce far greater magnifications.
By using this electron microscope, as it is called, it is
possible to distinguish objects one hundred times smaller
than the smallest object seen by the best light microscope,
for it is capable of magnifying up to × 200 000. But it has
disadvantages. Electrons are easily deflected from their
path, so objects being examined must be kept in a vacuum.
Usually, very thin specimens are examined *(figure 44)*, but
recently, a new modification, the scanning electron
microscope, has allowed three-dimensional studies as in
figure 46, showing the beautiful patterns on pollen grains.

Q1 Robert Hooke was the first person to use the word 'cell' for
the units of life. Why did he choose this word?

Q2 How did the 'cells' he drew in *figure 41* differ from those of
the broad bean root tip shown in *figure 35*?

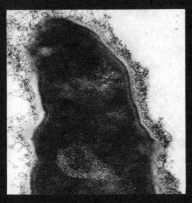

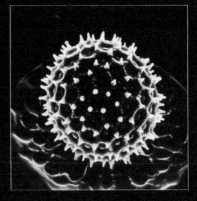

Figure 45
An electron microscope.
*Photograph, AEI Scientific
Apparatus Ltd.*

Figure 44 *(above)*
Electron micrograph of part of a
bacterium (×50 000).
*Photograph, Dr Audrey Glauert,
Strangeways Research Laboratory,
Cambridge.*

Figure 46 *(above)*
A pollen grain *(Ipomoea)* as seen
with a scanning electron
microscope (×310; photograph
taken at ×620).
*Photograph, Dr Patrick Echlin,
Department of Botany, University
of Cambridge.*

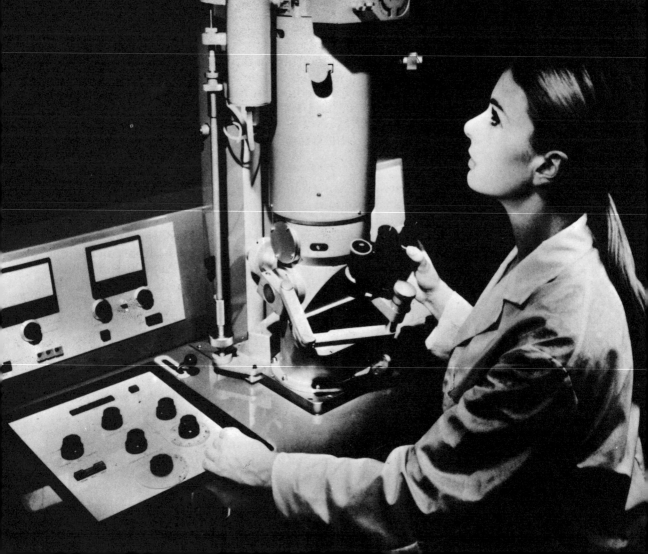

4

How living things begin

When you examine the root of a plant and the inside of the cheek as in Chapter 3, you see that both consist of cells. The bodies of all plants and animals are made up of cells, but these are not all alike as they have different and special jobs to do. Some cells, called the sex cells, have a special function when the animal or plant reproduces.

Sex cells are of two kinds. In animals these are called the *egg*, which is produced by the female, and the *sperm*, which is produced by the male.

Egg cells are among the largest cells and birds' eggs include examples of the largest single cells of all.

4.1 What is inside an egg?

The inside of a hen's egg is difficult to examine but you can see quite a lot by looking at a hard-boiled egg or a raw egg.

Examining a hard-boiled egg
1 Take off the shell carefully. Try to remove the ends in large pieces.
2 Cut the egg in two lengthwise on a tile. Compare your egg with others.

Q1 Is the yolk always in the middle of the white?

Q2 Is the yolk the same colour all through?

Examining a raw egg
1 Support the egg in a dish on some paper towelling. This will help to stop it from turning round.
2 Using the handle of a scalpel, crack the egg across the top and pick off the bits of broken shell until you have made a window in the shell as in *figure 47*. The window may also be made by cutting out a circle of shell with a pair of pointed scissors.

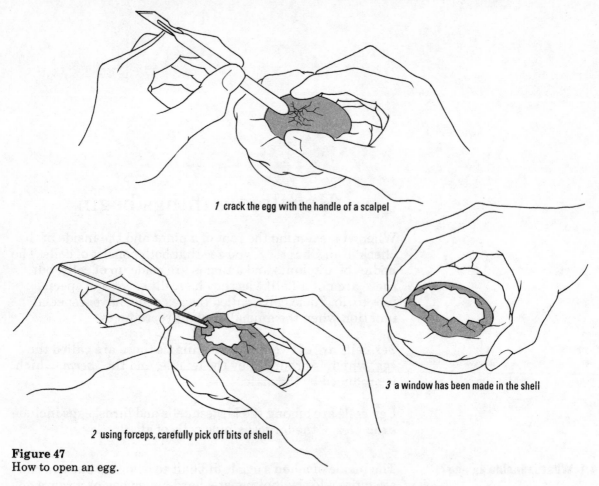

1 crack the egg with the handle of a scalpel

3 a window has been made in the shell

2 using forceps, carefully pick off bits of shell

Figure 47
How to open an egg.

Q3 Is there anything lying on the yolk?

3 Pour water into a dish and carefully tip the egg into it, if possible without damaging the yolk.

4 Look inside the broken shell.

Q4 What do you find just inside the shell of the egg?

5 Now examine the white.

Q5 Is the appearance of the white the same all through?

Look at *figure 48* and see if you can make out, on your egg, all the parts labelled in the diagram.

Q6 How much of the whole egg that you have examined is a single cell?

If a poultry farmer wants to produce eggs that will hatch into chickens, he must let his hens mate with a cock.

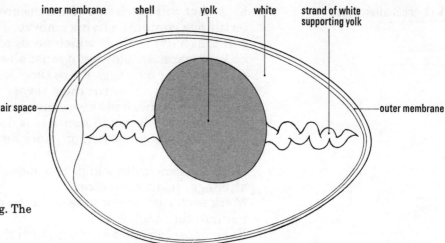

inner membrane   shell   yolk   white   strand of white supporting yolk

air space

outer membrane

Figure 48
The structure of a hen's egg. The egg has been cut into two, lengthwise.

Examining an egg laid by a hen which has been mated

1   Place the egg on crumpled paper towelling in a dish. If the egg has been in an incubator this will help to keep it warm as well as stop it from rolling about.

2   Cut a window in the egg in the same way as before.

Q7   How does this egg differ from the first raw egg you examined?

3   Pour warm salt solution into a dish and carefully tip the egg into it, if possible without damaging the yolk.

4   With a pipette draw up some warm salt solution and drop a little onto the yolk.

5   Under a lens or a binocular microscope you may be able to see more detail.

Figure 49
A cock mating with a hen.
After Guhl, A. M., 'The social order of chickens'. Copyright © 1956 by Scientific American, *Inc. All rights reserved.*

How living things begin

4.11 Incubating eggs

Eggs that will hatch into chickens are called fertile eggs. A fertile egg contains a living embryo. That is to say, each one contains a living chick which needs to respire, feed, and to be kept warm. In addition it must also be kept moist. A hen sitting on her own eggs keeps them warm (at a temperature of about 38 °C) and turns them several times a day. The moisture from her body keeps the eggs in a damp atmosphere. The incubator drawn in *figure 50* is designed to copy the conditions which the hen provides for hatching her eggs.

Here are some rules which you should follow when looking after eggs in an incubator:
1 Mark each egg, on one side, with the date on which it is put into the incubator.
2 Every morning and afternoon, turn the eggs by rolling them over sideways with the tips of the fingers. If the marked side was on top, roll the egg over so that it is now underneath. Wash your hands before turning the eggs.

Figure 50
Section through an egg incubator.
The arrows show how the air flows through it.

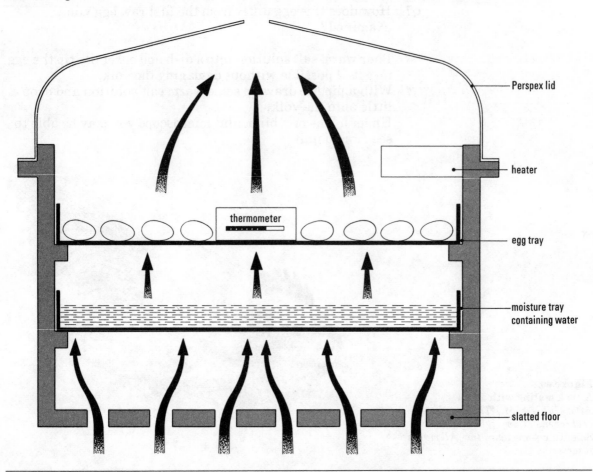

Perspex lid

heater

thermometer

egg tray

moisture tray containing water

slatted floor

From observations which you have made and from other knowledge which you now have, answer these questions:

Q1 Why should the eggs be turned at least twice a day?

Q2 Why should there be a current of fresh air passing through the incubator?

Q3 Why do the eggs need to be in a moist atmosphere?

Q4 When you take an egg from the incubator to examine the embryo, why should you be careful to hold it the same way up as it was on the egg tray?

Q5 Why is the incubator kept at 38 °C?

Q6 Why should you wrap the egg in paper tissue or cottonwool before transferring it to the place where it is going to be examined?

Fertile eggs can be kept and labelled in an incubator such as the one shown. The rate of growth of the embryo can be studied by removing and opening the eggs at intervals. But first of all, how does an egg cell start to grow into an embryo?

4.2 How does an egg cell start to grow into an embryo?

Figure 51
An egg cell is fertilized.
a Sperms surround an egg cell.
b The head of one sperm is about to join the nucleus of the egg cell.

If you could look at an egg inside a hen which had been mated to a cock, you might be able to see, with the aid of a microscope, what happened at the beginning to produce the embryo. The egg cell, like other cells, has protoplasm and a nucleus. Somewhere on the surface of the yolk there is a very tiny speck which shows the position of the nucleus. The yolk forms the rest of the egg cell. With higher

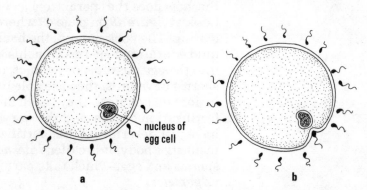

nucleus of
egg cell

a   b

magnification, you might see a large number of even smaller cells, swimming about near the egg cell, each with a tail which it uses for movement (see *figure 51*). These are called *sperms* and are produced by the cock. In *figure 51*, one of the sperm cells is pushing through the membrane surrounding the egg and eventually joins with the egg nucleus. By this action the egg is *fertilized*.

Once this has happened, the fertilized egg starts to divide – first into two cells, then into four, eight, sixteen, and so on until there is a group of cells which is the beginning of the embryo. *Figure 52* shows this process of cell division. The example used is the egg of a starfish because the cells are large and easy to study, but the same process takes place in a fertilized hen's egg, except that the cells form a flat circular sheet instead of a ball.

Figure 52
Four fertilized eggs of a starfish are dividing. The lefthand egg has divided into two cells. The two eggs in the centre have each divided into four cells. The egg on the right has divided to make a ball of cells.
Photograph, Hugh Spencer/Frank W. Lane.

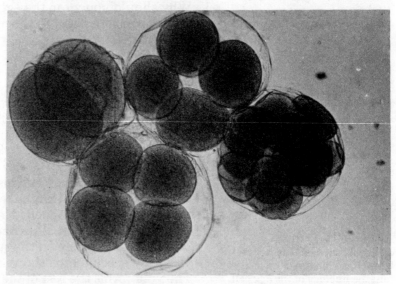

But how does the sperm from a cock reach the egg cell? Look at *figure 49* on page 49 where the cock and hen are mating. The cock mounts the back of the hen and passes fluid containing sperms from his sperm duct under the tail into the egg passage, or oviduct, of the hen. Because the sperms have tails, they are able to swim up the egg passage inside the hen until they reach an egg cell. There, fertilization occurs, and the egg then starts to divide as we have just described. When fertilization of the egg occurs inside the body, it is called *internal fertilization*. The sperms and eggs which take part in the process are known as *gametes*.

This kind of reproduction, in which a new generation is started by the union of the sperm (male gamete) with the egg (female gamete) to form the fertilized egg, is called *sexual reproduction*.

It is difficult to see fertilization in a hen because it takes place inside the female. Fertilization can be observed quite easily in the small marine worm *Pomatoceros triqueter*. This lives in a white tube of lime with a keel on top; it is often found firmly attached to rocks and pebbles between tide marks (see *figure 53*). It is possible to push the worms out of

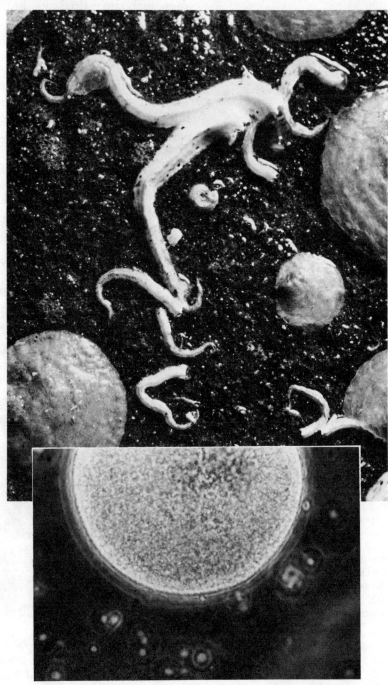

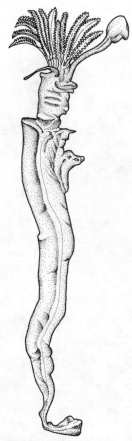

Figure 53 *(right)*
The white tubes of lime of the marine worm, *Pomatoceros triqueter* (seen here with saddle oysters, *Anomia ephippium*). *Photograph, Heather Angel.*

Figure 54
Pomatoceros triqueter in its tube.

Figure 55 *(right)*
A photomicrograph of the egg of *Pomatoceros* (100 μ in diameter), surrounded by sperms. The heads of two sperms have just pushed through the membrane surrounding the egg.
Photograph, Professor J. Cohen.

their tubes with a blunt seeker. You can tell the males from the females because their bodies are yellow, while the females are either red or violet in colour. As soon as the worms are removed from their tubes, the males release sperms and the females release eggs and fertilization takes place in the water. Both eggs and sperms are too small to be seen with the naked eye, but under the microscope they can be seen quite distinctly.

4.21 Different kinds of eggs and sperms

Figure 56 is a photograph of a sperm fertilizing a rat egg. The egg and sperm are magnified many hundreds of times but you can see that the egg cell is much larger than the sperm. Because the egg has no tail, it cannot move itself about.

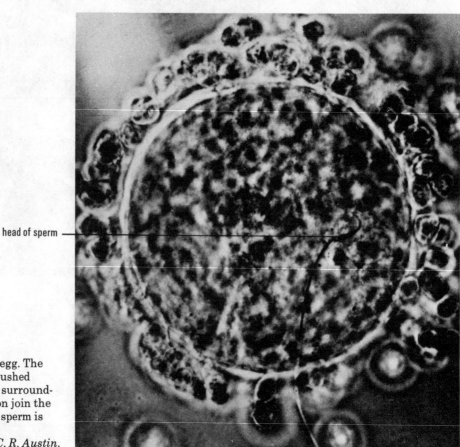

head of sperm

Figure 56
A sperm fertilizes a rat egg. The head of the sperm has pushed through the membrane surrounding the egg and will soon join the nucleus. The tail of the sperm is still outside the egg.
Photograph, Professor C. R. Austin.

Figure 57 shows the eggs of a number of different animals and gives details of their sizes. Some can be seen quite easily with the naked eye. For instance a hen's egg is relatively large and weighs about 65 g. The eggs of some birds are even larger – that of an ostrich is 18 cm long and

15 cm in diameter. Frog and fish eggs can be quite large too. All these contain a lot of yolk which nourishes the growing embryo. But the eggs of most mammals are microscopic, like that of a rat. This seems strange since mammals can grow so large; later you will discover why it is a good thing.

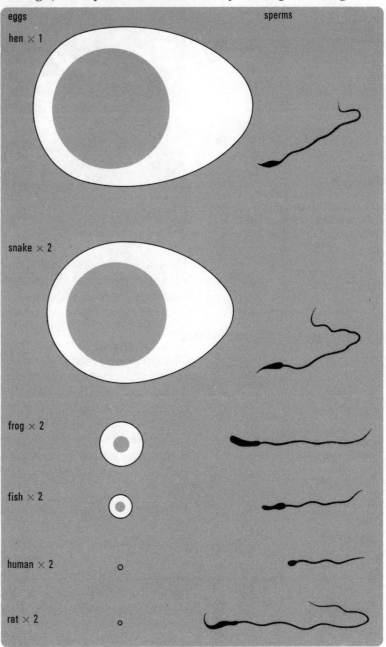

Figure 57
Eggs and sperms of different animals. The sizes of the eggs are given. Note that the sperms are all magnified approximately 450 times. In the hen, snake, frog, and fish the egg cells are surrounded by other materials such as white (albumen) and a shell.

Sperm cells are always far too small to be seen without a microscope. They vary in shape but their tails enable them to move at about a speed of up to 0.5 mm per second provided there is some kind of liquid in which to swim. Unlike the egg, they have practically no food store and thus cannot live for long. *Figure 58* shows the sperms of a boar. Notice the head of each sperm and its long tail.

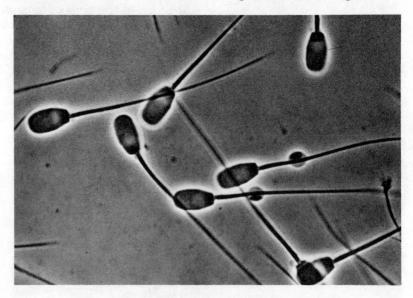

Figure 58
Sperms of a boar (× 1575). Like those of the rat (see *figure 56*), each of these has a head and a tail.
Photograph, Professor T. Mann.

Q1 How are animals' eggs and sperms different? Make a list of all the differences. Whenever possible explain the advantages or disadvantages that result.

4.3 How does the hen's egg develop?

The hen's egg starts to develop after it has been fertilized inside the egg passage of the hen. It begins to divide into a number of cells (just like the eggs of the starfish shown in *figure 52*). As the egg passes down the oviduct before being laid, it divides all the time to form the beginning of an embryo and the walls of the oviduct produce the white (*albumen*) which surrounds the yolk. A shell is added just before the egg is laid, when the embryo is already 20 hours old.

While the egg has been inside the hen, it has remained at a temperature of 38 °C. So if we want to allow the embryo to go on growing, we must put the newly laid egg into an incubator at the right temperature. The first raw egg which you examined in section 4.1 was laid by a hen which had not mated with a cock so the egg cell was not fertilized and we say it was *infertile*. The other raw egg, on the other hand, was the result of a cock mating with a hen. A fertilized egg can grow into an embryo chick.

Q1 What functions do each of the following parts of the egg have for a developing chick?
a the yolk
b the white
c the membranes and shell

4.31 The chick embryo grows

If fertile eggs are incubated it is possible to find out how quickly the embryo grows and to see how the embryo develops new structures as it gets older. This may be done by opening eggs which have been incubated for different periods of time or by studying photographs and films which show the different stages of development.

Look at the photograph and drawing of the 3½-day-old chick embryo (*figures 59* and *60*).

Figures 59 and 60
A 3½-day-old chick embryo (× 10).
Photograph,
Brian Bracegirdle.

Q2 Which features are already formed?

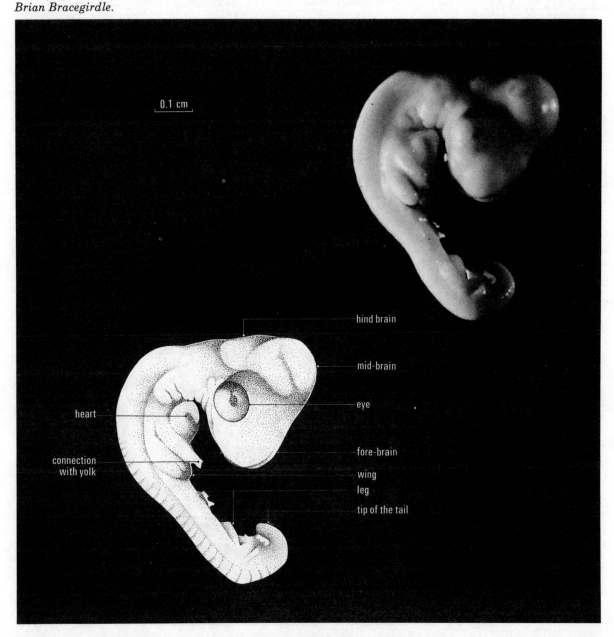

0.1 cm

hind brain

mid-brain

eye

fore-brain

wing

leg

tip of the tail

heart

connection
with yolk

Now look at the photograph and drawing of the 5-day-old chick embryo (*figures 61* and *62*).

Q3   Are there any new structures formed since the embryo was 3½ days old?

Q4   How does the size of the eye compare with the size of the head in the 3½-day-old embryo and the 5-day-old embryo?

Q5   Have any other structures changed since the chick embryo was 3½ days old?

Figures 61 and 62
A 5-day-old chick embryo (×12).
*Photograph,
Brian Bracegirdle.*

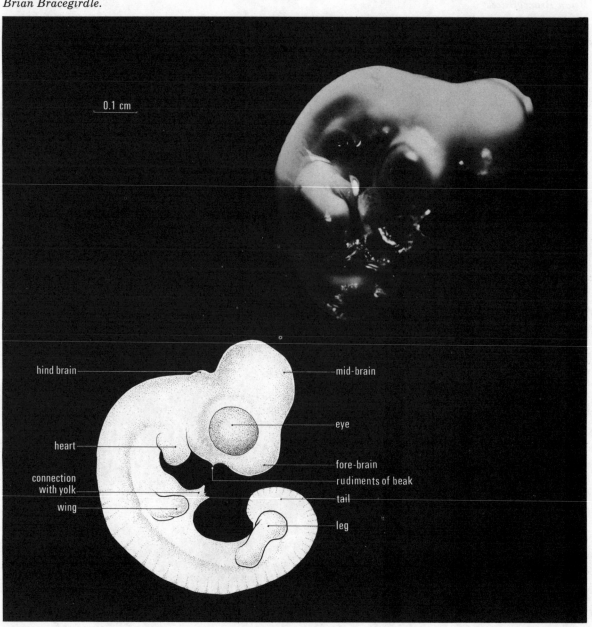

0.1 cm

hind brain
mid-brain
eye
heart
fore-brain
rudiments of beak
connection with yolk
tail
wing
leg

Look at the 7-day-old chick embryo (*figures 63* and *64*).

Q6 What changes occur in the structures of the chick embryo between the fifth and the seventh day of its development?

Figures 63 and 64
A 7-day-old chick embryo (× 4.0).
Photograph, Brian Bracegirdle.

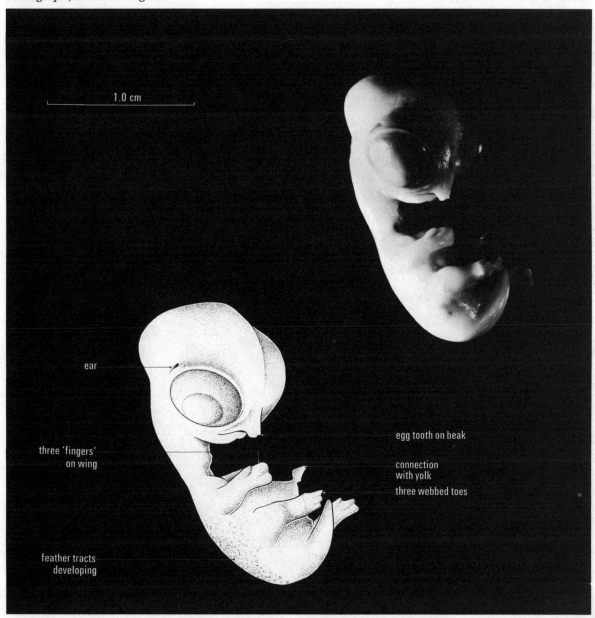

ear

three 'fingers' on wing

feather tracts developing

egg tooth on beak

connection with yolk

three webbed toes

Now look at the 10-day-old chick embryo (*figures 65* and *66*).

Q7   How does the size of the eye compare with the size of the head now?

Q8   What new structures do you see at 10 days old?

Q9   How have other structures developed since the embryo was 7 days old?

Figures 65 and 66
A 10-day-old chick embryo (×2.25).
Photograph, Brian Bracegirdle.

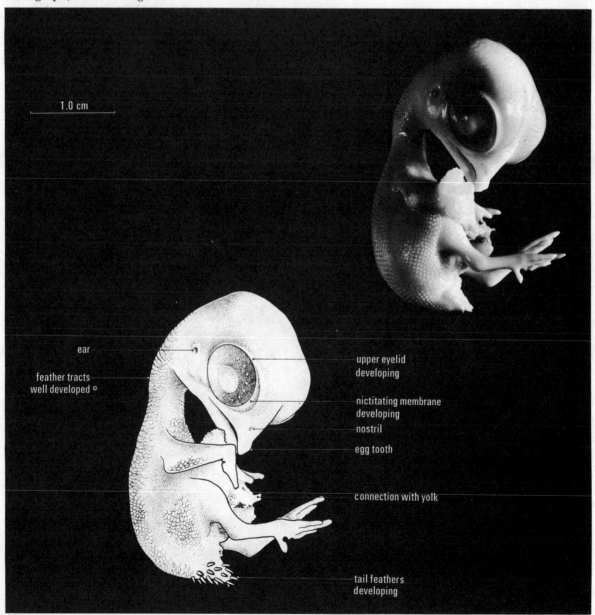

1.0 cm

ear

feather tracts
well developed

upper eyelid
developing

nictitating membrane
developing

nostril

egg tooth

connection with yolk

tail feathers
developing

Figures 67 and *68* show you the chick embryo at 15 days and at 19 days.

Figure 67
A 15-day-old chick embryo (× 1.5).
Photograph, Heather Angel.

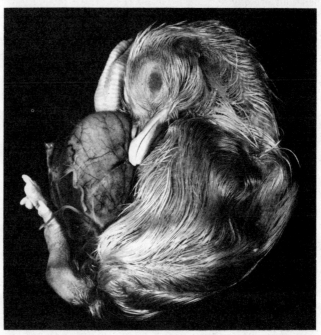

Figure 68
A 19-day-old chick embryo (× 2).
Photograph, Heather Angel.

Q10   What developments take place between 7 days old and 19 days old?

Most birds have a *nictitating membrane* over each eye. It is a third eyelid which protects the eye of the bird from dust and the glare of the sun when it is in flight. Look at your neighbour's eye. Right in the corner, can you see a small piece of pink skin? This is all that remains of the nictitating membrane in human beings.

Scientists investigating the way chick embryos grow might need to measure their growth in some way. Some of the questions to be considered when deciding the best way to do this are:
Would it be *a* easier or *b* more accurate, to measure the mass or length of the embryo?
How could you determine the mass of a very small embryo accurately?

The mass has been measured as you can see from *table 6*, but it would be difficult for you to do.

Days	Grammes
3	0.02
5	0.13
7	0.57
10	2.26
12	5.07
14	9.74
16	15.98
18	21.83
20	30.21

Table 6
The mass of chick embryos from the third to the twentieth day.

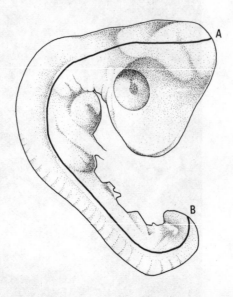

Figure 69
How to take the measurements of the length of the chick embryo.

Measure the length of the embryo by laying a piece of string from A to B, following the area of the backbone as shown.

Introducing living things

Plot the figures in *table 6* on a graph and see what it tells you about the chick's growth.

But you can actually measure the length of the embryo as it grows, or do it by referring to the photographs and drawings. *Figure 69* shows you how to take the measurements from the photographs, using a piece of string, which you then measure with a ruler.

In each drawing you are shown the distance which represents a centimetre so you can work out the actual length of the embryo. Do this up to the tenth day. All the information you collect can then be put together in a table similar to that shown in *table 7*.

Plot the growth in length of a chick embryo on a bar graph.

Age of embryo	Length of embryo	Notes
$3\frac{1}{2}$ days	12 mm	Head very large, etc.
5 days		
7 days		
10 days		

Table 7

Q11 Join the bars with a line running through the middle of the top of each bar. If it is *a* straight, or *b* curved, what does your graph tell you about the growth in length of the embryo chick from the $3\frac{1}{2}$-day stage to the 10-day stage?

Q12 Compare the graph of length with the one you plotted of the mass of the chick embryo from *table 6*. What do they tell you about the rate of growth of chick embryos?

The photographs and drawings of the different stages of the chick embryo's development inside the egg do not give any information about what is happening to the yolk or about other structures such as blood vessels or membranes. For this it is necessary to open an egg at the different stages and look. Alternatively, it may be shown on a film.

4.32 The chick hatches

It may be possible for you to incubate eggs until they hatch. They should start to hatch on the twenty-first day and the actual process of hatching is most interesting to watch. Movements of the chick within the shell cause the head to be jerked forward and the beak upwards. The egg tooth on the beak is now quite large and these movements split the membranes inside the shell and finally crack the shell itself. The chick kicks with its legs and the beak enlarges the hole

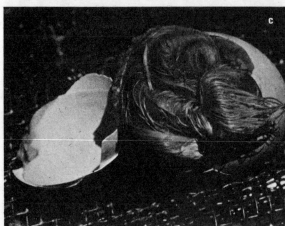

Figure 70
A chicken hatches from an egg.
a The membranes inside the egg are split and the shell cracked.
b The beak of the chicken enlarges the hole in the shell.
c The damp chick is still curled up but is nearly free.
d The chick has hatched and its feathers are now dry and fluffy.
Photographs, Pictorial Press Ltd.

in the shell. When it first appears the chick looks rather shaggy and damp but its downy feathers soon dry. It is amazing to think that the single-celled embryo, with which we started the story, can develop, in just 21 days, into a day-old chick covered with feathers and able to peck food.

Write your own account of a chick hatching. It will be best of all if you watched it happening, but otherwise, base your account on the photographs or a film.

All things from the egg?

In this chapter you have studied the development of a hen's egg into a chicken. We know many other examples of animals which are formed from eggs. *Figure 82* on page 82 shows a young crocodile hatching from an egg. Things like this were easy for our ancestors to observe but there are many facts about reproduction which we take for granted and which were much less easy for them to understand.

The ancient Greeks were interested in many aspects of natural history and some of their ideas were very sound and based on careful observation. What they observed, however, did not always give them correct answers. The Greek philosopher Aristotle thought that eels arose from the mud at the bottom of ponds. The books of the ancient Greek scientists were later translated into Arabic and then into Latin and their beliefs survived until the Middle Ages and even more recently.

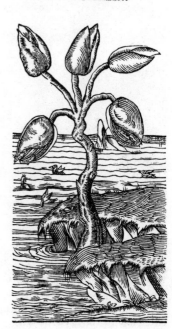

Britannica Concha anatifera.
The breede of Barnakles.

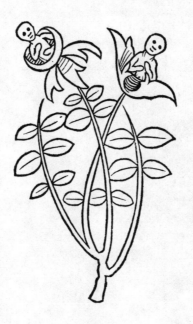

Figure 71 *(left)*
From Hortus sanitatis *(1491).*
By courtesy of the Royal Botanic Gardens, Kew.

Figure 72 *(right)*
The goose tree.
From The Herbal, or Generall Historie of Plantes &c., *by John Gerard, 1597 (1636 edition). By courtesy of the Royal Botanic Gardens, Kew.*

Gradually, scientific experiment, with properly designed controls, sifted some of the facts from the fallacies. William Harvey (1578 to 1657), the discoverer of the circulation of the blood, was the first man to appreciate that mammals developed from eggs and he was bold enough to suggest that everything grows from an egg. He was physician to Charles I who allowed him to experiment on the deer in the royal parks and it was from his dissections of the deer that he drew his ideas.

Later it was understood that before the egg could develop into a new animal and be born, it had to be fertilized. This is true of most animals though there are exceptions. Our familiar stick insects are nearly always female (male stick insects are very rare) and yet they lay eggs which hatch into young. At certain times of the year, other insects, too, can produce young without mating; an example is the

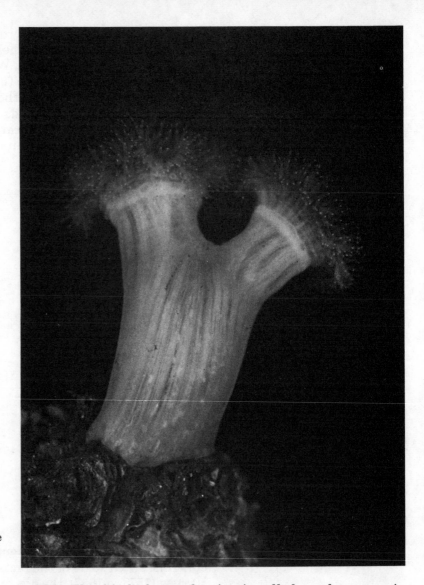

Figure 73
The sea anemone *(Metridium)* gives an example of asexual reproduction. Here it is dividing. After a time the division will extend down to the base. Then the two anemones will separate.
Photograph, Dr. D. P. Wilson.

aphids. This kind of reproduction is called *parthenogenesis*, a Greek word meaning 'virgin birth'. It is dealt with again in Chapter 9.

But besides reproduction involving the development of eggs there is also a form of reproduction which involves one organism only and in which no gametes are produced. This is called *asexual*, meaning without sex. One example is the famous one-celled animal *Amoeba* which simply divides its body into two. Other Protozoa may split into two or many more 'daughters'. Some members of the jelly-fish family can also reproduce in a similar way. *Figure 73* shows one sea anemone dividing into two and *figure 74* shows a small freshwater member of the same family, *Hydra*, producing daughters by *budding*. A small group of cells in the side of the animal begins to divide and bulge out. This 'bud'

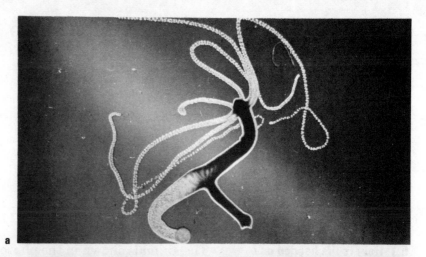

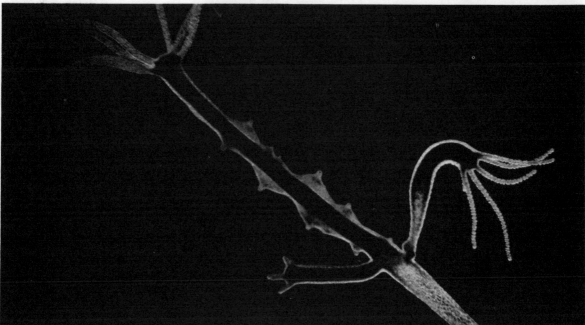

Figure 74
Here you can see two stages in the production of a new *Hydra* by budding.
a One bud has formed on the side of the parent.
b The first bud has now got tentacles and a second bud has formed.
This animal also reproduces sexually. Above the two buds, you can see sexual organs forming, each producing sperms or an egg.
Photograph, P. S. Tice.

gradually develops tentacles and a mouth, the part connecting the animal and the bud wears through, and the daughter falls away to lead an independent life. Plants can also be made to reproduce asexually when gardeners make them increase by taking cuttings of their roots or stems instead of sowing seeds. You can read more about this in Chapter 6.

William Harvey was not quite right when he said that every living thing develops from an egg though it certainly applies to most animals. His idea helped the work of many later scientists, including Pasteur whom you will read about in Chapter 7. Pasteur showed that all cells come from other cells and, therefore, all living things can only develop from other living organisms of the same kind.

How living things develop

5.1 How is life handed on?

The animals shown in the photographs in *figure 75* are illustrating certain aspects of behaviour which have a connection with how life is handed on.

b

a

Figure 75
a Rhinoceroses.
Photograph, Ahni Productions, Tokyo.
b Peacocks.
Photograph, Barnaby's.
c Common toads.
Photograph, Heather Angel.
d Acorn barnacles.
Photograph, Heather Angel.
e Dogfish.
Photograph, Dr Douglas P. Wilson.
f Sticklebacks.
Photograph, Dr Douglas P. Wilson.

Q1 Describe what each of the animals is shown doing in
figure 75. You may also look at the frogs and the African
clawed toads in figure 79 (page 74), the domestic fowl in
figure 49 (page 49), and the locusts shown in figure 189
(page 212).

Q2 In each case what would you suggest is the purpose of the
activity shown?

Q3   Is there any means of distinguishing which is a male and which is a female in each picture shown and, if so, what is the difference?

Q4   Which of the animals in these figures
  a   lay eggs on land
  b   lay eggs in water
  c   lay eggs with hard shells
  d   do not lay eggs?

Q5   Summarize the main facts about sexual reproduction in animals by answering the following questions.
  a   *What* is needed if animals are to reproduce sexually?
  b   *Where* are animals able to reproduce?
  c   What sort of *behaviour* is connected with animals reproducing?
  d   What is the *advantage* of this behaviour?

5.11 How to tell the sex of animals   If you keep any small mammals as pets at home it is very useful to be able to tell a male from a female. This can be quite difficult in young mammals but the difference becomes more obvious as they get older. When sexing mammals it is always very important to handle them correctly. *Figures 76* and *77* show you how to do this.

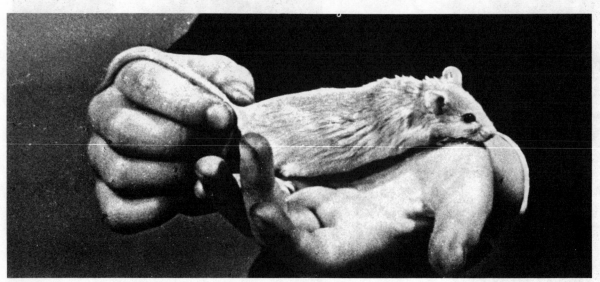

Figure 76
The right way to pick up and hold a mouse.
By courtesy of Schools Council publications; photograph supplied by John Wray of the Educational Use of Living Organisms Project.

Handling mice
Always pick a mouse up by the base of its tail near the rump, as shown in *figure 76*. The animal must not be held in this way for any length of time, however, without its body being supported on a firm surface. Mice soon become used to being handled, but you must be gentle and not squeeze the mouse if it struggles.

Introducing living things

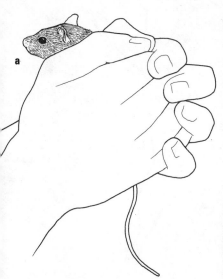

Handling gerbils

The technique recommended for picking up mice should not be used on animals with furry tails like gerbils as there is a possibility of fur stripping from the tail. Gerbils should be 'cupped' in the hands (*figure 77a*) and this is also a useful method for handling other small animals such as hamsters. The animal is stroked and allowed to run over the hands. It is lifted with a scooping motion, using both hands, which form a cup. The fingers and thumbs may be used to restrain the animal, which should have its head pointed towards its handler. *Then*, in order to tell its sex, you may hold the gerbil up very gently by the base of the tail and turn it as in *figure 77b*.

Figure 77
a Holding a gerbil by cupping.
b Telling a gerbil's sex.
Photographs, J. R. Lance.

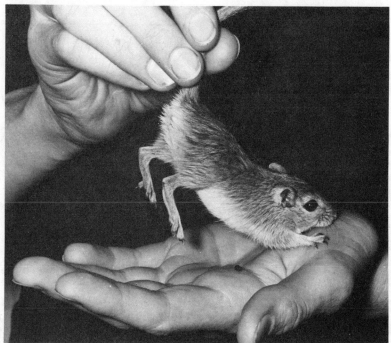

Sexing mice

The drawings in *figure 78* show how you can tell the difference between a male and a female mouse. The most obvious difference is in the distance from the genital papilla to the anus. In the adult male this is about 12 mm while in the female it is only about 6 mm. However, with live specimens, these differences may be less clear. Only by examining several mice can you be certain of the sex of each because it is necessary to compare the distances in the two sexes. In the adult female the teats of the mammary glands should be visible.

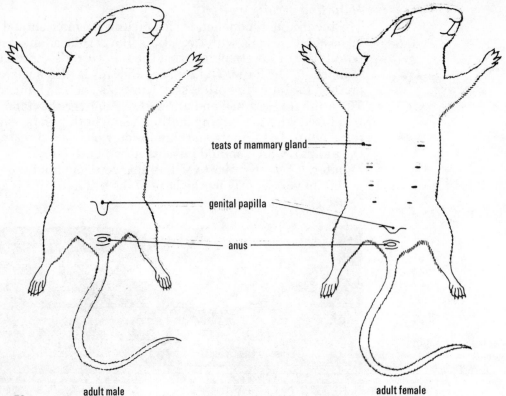

teats of mammary gland

genital papilla

anus

adult male

adult female

Figure 78
How to sex mice.

You will probably need help with sexing young mice.

5.2 How do animals breed in water?

In animals such as birds the eggs are fertilized by the sperms inside the body of the female and so fertilization is internal. But in the marine worm, *Pomatoceros*, the sperms released by the male swim in the sea water to reach the eggs released by the female. When fertilization takes place outside the body in this way, the process is called *external fertilization*.

The eggs of many species of freshwater and marine animals are fertilized externally and the sperms swim, by means of their tails, to reach the eggs.

Herrings live in shoals, and at certain times of the year, the males and females congregate in huge numbers on the sea bed. As soon as the eggs are laid they sink to the bottom and the male herrings release sperms into the water around them. The sperms swim in the sea and eventually some of the eggs will be fertilized. But, as you can imagine, this is a very wasteful process. One female herring lays as many as 10 000 eggs in a year while the male produces many

Introducing living things

millions of sperms. Some eggs may not be fertilized and therefore will not develop, while millions of sperms will also die. Thus for aquatic animals, external fertilization is a risky process involving the production of large numbers of eggs and sperms in order to make sure that at least a few eggs will be fertilized.

We cannot watch the process of fertilization in a fish living in the sea, but some of the easiest animals to study are frogs or toads. During the spring they leave their winter quarters – under vegetation or in damp ditches – and make their way to ponds and slow-flowing streams. They may have to travel overland for a mile or more to find water. The males often get there first, and await the arrival of the females. A male will then get onto the back of a female and hold her firmly with his front legs around her armpits.

Soon the female will start to lay eggs, and as the eggs are laid the male pours sperms over them in the water. We cannot see the sperms because they are far too small, but like all sperms they swim through the water by means of tails and so reach the egg. Several hundred eggs may be laid by one frog, but not all will develop into frogs.

Q1 Can you suggest reasons why?

5.21 The African clawed toad

Our native frogs and toads reproduce in the springtime. Many children have collected frog spawn from ponds and ditches in order to take it home or to school to watch the tadpoles hatch and the tiny frogs develop. Unfortunately, many of the eggs collected were never reared to become frogs. Unfortunately, too, frogs are no longer found in many of the places where they were once quite common and the number of frogs in the country has been slowly getting less in recent years. Therefore we should not take large amounts of frog spawn out of ponds but should leave it to develop in the hope that the number of frogs will increase again.

If you take a small amount of frog spawn, you should always release the tadpoles or tiny frogs, if possible at the place where they were originally collected.

Q2 What do you suggest may have contributed to this recent decrease in the number of frogs?

If we wish to avoid collecting our native frog spawn or if we wish to study reproduction at other times of the year we can make use of the African clawed toad, *Xenopus laevis*.

As the name suggests, these animals come from Africa. They are mostly found in Cape Province in the south, but also occur in many parts of East Africa. (*Figure 79a*.)

Our native frogs and toads are ideal for the study of mating, egg-laying, and development of the tadpoles, but details of their life history have not been given, since you can easily find a good account in almost any natural history book. The stages in development of *Xenopus*, however, will be unfamiliar; therefore, these are described in some detail in the following sections.

Figure 79
a African clawed toads (*Xenopus laevis*) mating.
Photograph, J. R. Lance.
b Frogs (*Rana temporaria*) mating.
Photograph, Heather Angel.

Introducing living things

5.22 Mating and egg-laying in the African clawed toad

African clawed toads will not usually mate and lay eggs in the laboratory of their own accord. But we can encourage them to do so by injecting into both a male and a female toad a substance which does not harm them, but which starts off the mating process. After the injections they are put into a small tank of water kept at 22°C.

If your school keeps African clawed toads you may be able to see a pair of toads which are mating after they have been injected. Whether you study the African clawed toad or our native frog in detail it is interesting to observe both animals and make comparisons.

Q3 How does mating in the African clawed toad compare with the way the common frog mates?

Q4 What differences do you see between the male and female frog and the male and female African clawed toad?

Look at *figure 79* if you do not have the animals to observe at first hand.

After the African clawed toads have paired, the female will start to lay eggs and she will continue to lay for several hours. Under natural conditions a female will lay as many as 15 000 eggs, but in a small tank in a laboratory she may only produce a hundred or so.

If the eggs of both the African clawed toad and our native frog are available compare the eggs – for size and the amount of jelly surrounding them.

5.23 Examining the eggs of the African clawed toad (or native frog or toad)

Figure 80
Stages in the development of the egg of the African clawed toad *(Xenopus laevis).*

You can examine the eggs of the African clawed toad or frog in closer detail by transferring one or two eggs, with a pipette, into a watch glass, adding a little water, and looking at them with a hand lens or a binocular dissecting microscope. Draw what you are able to see.

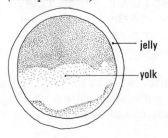

1 An egg which is about 1½ hours old

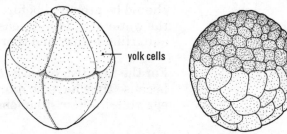

2 An egg which is about 2½ hours old   *3* An egg which is about 3½ hours old

Compare the drawings you have made with the drawings in *figure 80.*

Q5 Did the egg you looked at resemble any of the stages shown and if so, which stage was it at?

Q6 What must have happened to the egg for it to change from stage *1* to stage *2*?

Q7 What does it mean if an egg does not change from the stage shown in drawing *1*?

Q8 Describe in your own words what is shown happening to the egg as it changes from stage *1* to stage *3*.

Q9 Has the egg begun to grow (that is, add new material) when it is $3\frac{1}{2}$ hours old?

Q10 What effect does the yolk have on the division of the fertilized egg?

Q11 Which part of the egg do you think becomes the embryo?

5.24 The eggs hatch into tadpoles

If you are able to look at the eggs every day, you may notice that after about two days the inside of the egg seems to change shape and become longer. The young (embryo) tadpole is now visible and by the third day it may be ready to hatch, and you can watch how tadpoles are trying to struggle out of the jelly.

Very soon after they hatch the tadpoles will start clinging to the sides of the aquarium, hanging tail downwards. At this stage they have tiny frilled gills on either side of the head and breathe like a fish, taking in the oxygen dissolved in the water by means of these gills.

Some of the eggs in the aquarium may never hatch. You should already have discovered the reason for this. If these were left in the water they would decompose, the water would smell, and the young tadpoles would die. Instead of removing the eggs from stale water it is easier to take the tadpoles out and place them in a fresh tank of water. This should be prepared beforehand to receive the tadpoles – let the water stand for at least twenty-four hours and, using an aquarium heater and thermostat, keep it at 22°C to 23°C.

For the first two or three days the tadpoles cannot feed because they have no mouths but there is still quite a lot of egg yolk in their stomachs which they use as food.

After four or five days the tadpoles develop mouths and can feed. In the ponds and ditches of Africa in which they normally live, they eat microscopic green plants in the water. In the laboratory it is easier to feed them by using a substitute for microscopic green plants, such as ground-up dried nettle leaves which look like green powder. It is best to do this by placing about a teaspoonful of the nettle powder in a piece of nylon (a nylon stocking would do) about 15 cm × 15 cm. Make this into a bag with your fingers and soak it in the aquarium water. The nettle powder soon becomes moist and if you draw the bag to and fro through the water, the fine green particles will quickly disperse. Watch the behaviour of the tadpoles as you do this. How do they react? Do they remain clinging to the sides of the aquarium or do they swim about?

5.25 The tadpoles grow and develop

If you keep the water in the tank at 23 °C, you will be surprised how quickly the tadpoles grow in length and develop new parts to their bodies.

Figure 81 shows drawings of five stages in the development of the African clawed toad which you should watch for in any tadpoles developing in the laboratory. The drawings are not, however, placed in the order in which the tadpoles develop.

Q12 Study the drawings and decide what you think must be the order. Give reasons for the order you select.

The written account that begins on page 79 is in the correct sequence and gives more detail about the development of the tadpoles.

Q13 After studying the written account relate the different stages described to the drawings and check that you decided on the correct order when you answered question 12.

Q14 To make a summary of the development of African clawed toad tadpoles, make a larger copy of *table 8* in your notebook and complete it. (The table will need a double page of your notebook.) If you follow the development of our common toad or frog you could still complete a similar table to give you a summary.

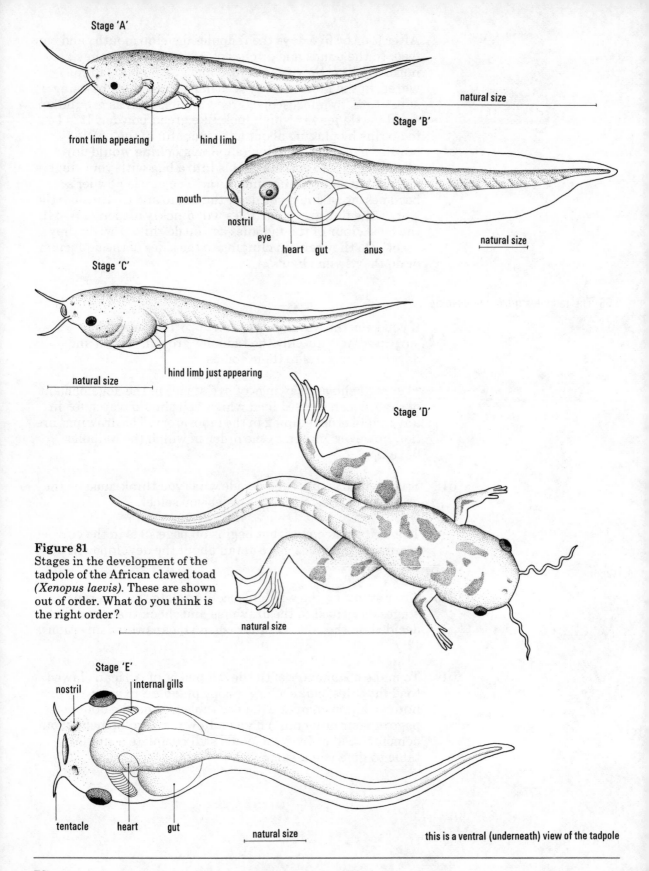

Stage 'A'

front limb appearing

hind limb

natural size

Stage 'B'

mouth

nostril

eye

heart

gut

anus

natural size

Stage 'C'

hind limb just appearing

natural size

Stage 'D'

Figure 81
Stages in the development of the
tadpole of the African clawed toad
(Xenopus laevis). These are shown
out of order. What do you think is
the right order?

natural size

Stage 'E'

nostril

internal gills

tentacle

heart

gut

natural size

this is a ventral (underneath) view of the tadpole

Drawings of stages (or refer to these by letters A, B, etc.)	How many days old it is	How long it is	What it feeds on	Where it gets its oxygen	What structure it uses for breathing	What part of the body it uses for moving about

Table 8
A table to contain a summary of the development of the African clawed toad (or the common toad or frog).

Stages in the development of the African clawed toad, *Xenopus laevis*

Check the following descriptions with your own observations of *Xenopus* in your school and make sure you record any differences you notice.

Stage 1 The tadpoles start to breathe air
At about four or five days old, the tadpole has just lost its external gills and its mouth is developing. When this happens it is growing lungs and starts to come up to the surface to gulp air. The tadpoles swim about the tank with darting movements, or hang head downwards with their tails flickering. In this position a tadpole is drawing in the particles of food through its mouth and sieving it out of the water through slits on either side of the head.

Stage 2 The tadpole develops tentacles
When the tadpole is about eight or ten days old there are two small tentacles developing at the front of the head. The mouth has got larger and you can make out the eyes and nostrils. The tadpole gulps water containing food. This is sieved out of the water as it passes over the sieves on either side of the inside of the mouth. Under a binocular microscope the sieving apparatus can be seen through the transparent skin, as well as the nettle powder which is circulating in the gut. The blood flowing through the heart can also be seen.

Stage 3 The tadpole develops back legs
When the tadpole is about twelve to eighteen days old, a small bulge appears at the back of each side of its head.

These bulges are the back legs starting to develop. The tentacles have also grown longer.

Stage 4 The tadpole develops front legs
At about five or six weeks old the tadpole has grown very large (it may even reach 80 mm in length). The tentacles are very long and the front legs, with four fingers showing, can be seen on either side of the head. The back legs are now well developed, with webbing between the toes. The tadpole now swims about using its webbed feet as well as its tail. The tadpole will not now grow any more in length.

Stage 5 The tadpoles change into toads
When the tadpoles are about seven or eight weeks old you may notice some strange things happening. They are now quite large (up to 80 mm long) and are beginning to look like little toads although they still have tails. At this stage in the laboratory you should keep the water in the tank fairly shallow so that the tadpoles can easily reach the surface to obtain air. The tentacles curl up and start to shrivel and the skin looks more like that of the adult toad. The mouth is wider and the tail shorter. The front of the head has changed shape to look more like that of the adult. The hind legs are now used for swimming.

A few days after these changes the tadpoles will have lost their tails altogether and become little toads. This rapid change from one form to another – from tadpole to toad – is called *metamorphosis*. Such changes from tadpole to adult occur in all kinds of frogs and toads, and in newts as well. Insects also change suddenly from one form to another. For instance, a butterfly changes from caterpillar to *chrysalis (pupa)* and from chrysalis to adult. Such changes are also examples of metamorphosis. (See Chapter 9.)

During metamorphosis the tadpoles stop feeding altogether but after they become little toads, they will no longer feed on nettle powder but need a meat, or carnivorous, diet. As the tadpoles change into toads they should be removed to another tank and must be fed at first on small living animals such as water fleas (*Daphnia*), blood worms (*Tubifex*), or the small white worms from compost heaps (enchytraeids). Soon they can be given shreds of raw sheep's heart, liver, or small earthworms.

Introducing living things

5.26 Comparing the development of *Xenopus* tadpoles with that of the common toad, *Bufo bufo*

If you have followed the development of common toad spawn you may be able to compare the development of the tadpoles with those of *Xenopus*. If so, try to answer these questions:

a   How long do tadpoles of the common toad keep their fringed gills?

b   After they have lost these external gills how do they breathe? Do they come up to the surface and gulp air like the *Xenopus* tadpoles?

c   Do tadpoles of the common toad have tentacles on the front of their heads?

d   How old are they
1   when they start to grow their back legs,
2   when they start to grow their front legs,
3   when they metamorphose?

5.3 How do animals breed on land?

You have been studying the development of eggs laid by *Xenopus*, or our own common frog (*Rana temporaria*), or toad (*Bufo bufo*). They lay their eggs in the water, as do most fish and some of the animals without backbones. The eggs develop outside the body in the water. Other animals lay eggs but not in water. For example, most reptiles lay eggs (and do so on land). In *figure 82* you can see a young crocodile just hatching. *Figure 83* shows the eggs of another reptile. All birds lay eggs and Chapter 4 deals with an embryo chick growing inside an egg. *Figure 84* shows a herring gull with its eggs, which are laid on the ground.

Q1   What difference, apart from size, is there between the eggs of fish and amphibians which are laid in water and the eggs of reptiles and birds which are laid on land?

Q2   What do you suggest is the reason for the difference?

At the beginning of the chapter you were asked questions about the activities of the animals shown in *figure 75* on pages 68 and 69 and those in *figures 49* (page 49), *79* (page 74), and *189* (page 212).

Q3   Look again at these figures and make two lists, one of those animals shown which breed in water and the other of those which breed on land. Then beside each one, write whether fertilization is internal or external. What statement might you now make about the fertilization of the eggs of land animals?

When a male animal introduces sperms into a female's body, this is called *copulation*.

Q4   What reason would you give for the fact that all land animals copulate?

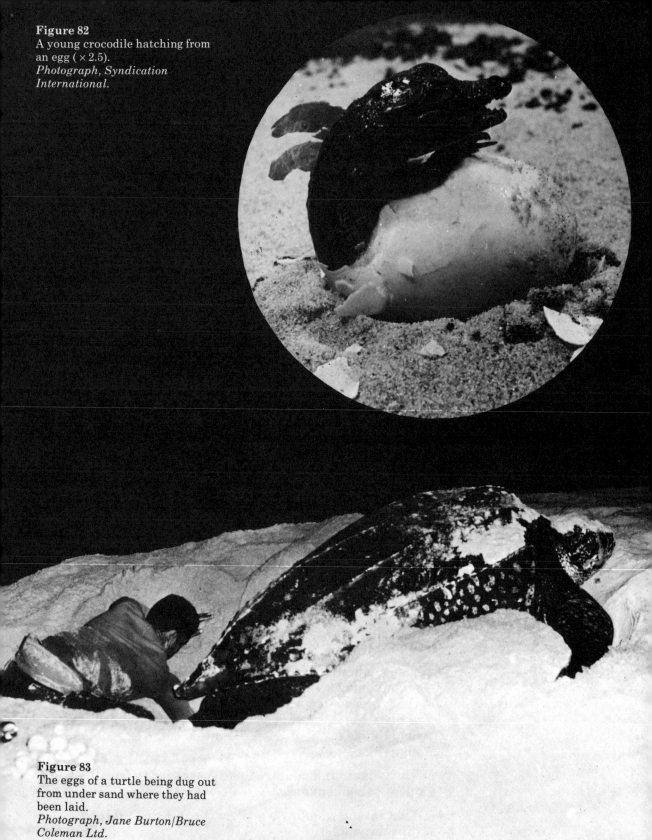

Figure 82
A young crocodile hatching from
an egg (× 2.5).
*Photograph, Syndication
International.*

Figure 83
The eggs of a turtle being dug out
from under sand where they had
been laid.
*Photograph, Jane Burton/Bruce
Coleman Ltd.*

Figure 84
A herring gull beside its eggs,
which it lays on the ground.
Photograph, Heather Angel.

5.4 How do mammals reproduce?

The mammals, that is, the group of animals to which we belong, do not lay eggs; instead, the embryos develop in a special place inside the body of the mother. When the young mammal has reached a certain stage of development it is born and starts to live outside the body of the mother.

5.41 The reproductive organs of the mammal

The characteristics of the different groups of animals are dealt with in detail in Chapter 10 and you will see that two of the special features of the mammal group are that their young are born and that the young get their first food from mammary glands. The rat is a typical mammal.

Look at the drawings of female and male rats in *figures 85* and *86*, dissected to show the parts of the body which are concerned with producing baby rats. These special parts are called *reproductive organs*. If you look at dissections of rats compare the drawings with the dissected rats and see if you can find the parts labelled in the drawings.

Figure 85
A female rat dissected to show the reproductive organs. The left uterus has been opened to show the embryos.

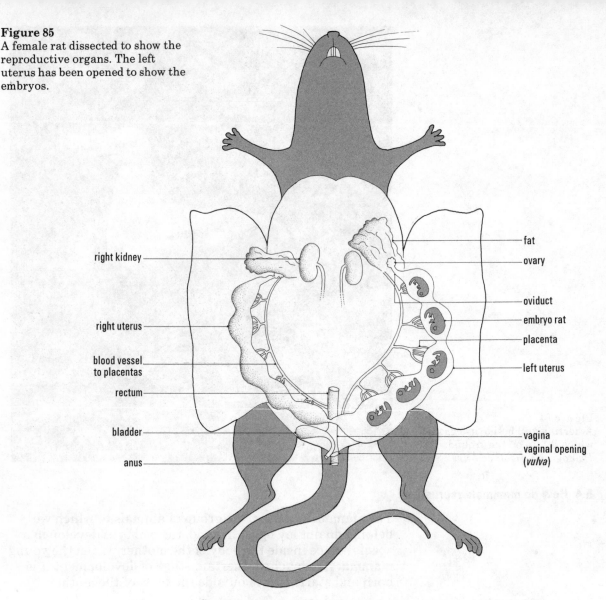

right kidney

right uterus

blood vessel
to placentas

rectum

bladder

anus

fat

ovary

oviduct

embryo rat

placenta

left uterus

vagina
vaginal opening
(*vulva*)

In the male rat, the reproductive organs consist of the *testes* (one testis on each side) in which the sperms are made, and a tube called the *vas deferens*, leading from each testis to the *penis*. The sperms travel along this tube.

In the female rat the reproductive organs look quite different. There is an *ovary* on each side of the body but it is small and rather difficult to find. The eggs, which are formed in the ovary, are only about 0.1 mm in diameter – far too small to see with the naked eye. You can see just how small a rat egg is compared with a hen's egg, if you look at *figure 57*.

As they are formed, the eggs are shed from the ovary and travel down a small tube, the *oviduct*, to the *uterus*. The

Introducing living things

Figure 86
A male rat dissected to show the reproductive organs. The left scrotum has been opened to show the testis.

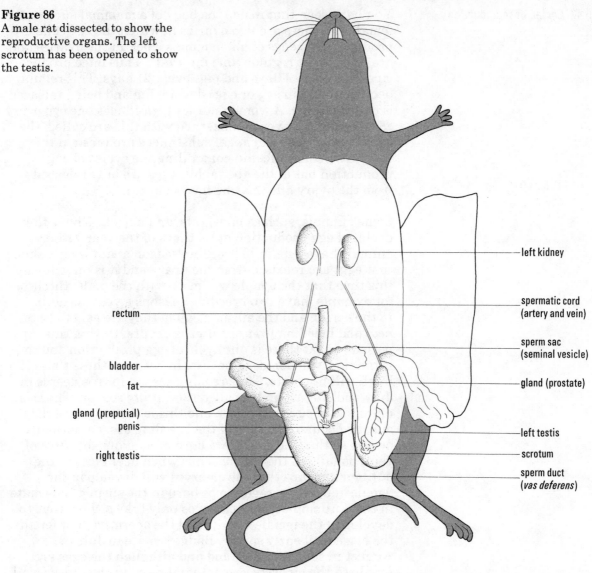

right uterus in the drawing in *figure 85* has bulges all along it. If you look at the left uterus in the drawing of the dissected rat, which has been cut open, you may be able to see that each bulge of this kind is really a baby rat. Each embryo is attached to the wall of the uterus by blood vessels and other tissues which together form what is called the *placenta*. Each embryo lies in a sac (the *amnion*) containing fluid. When they are ready to be born, the baby rats leave each uterus, one at a time, travel down the single *vagina*, and are born into the outside world through the vaginal opening.

Q1 Can you suggest, from looking at *figure 85*, how many young a rat might be likely to have in one litter?

You have seen that inside the body of a mammal such as the female rat, there are two ovaries in which the egg cells are produced. The eggs of all mammals ripen in the ovaries and are released at regular time intervals. Thus mice produce ripe eggs every 4 days and pigs every 21 days. This regular occurrence of an egg or eggs developing and being released is called a cycle. A woman normally produces one egg every 28 days and the events concerned with this are called 'the menstrual cycle'. Chemical substances produced in the body have some effect in controlling the cycles of egg production but in the doe rabbit, eggs are only released from the ovary *after* mating has occurred.

Some animals, such as mice, rats, and hamsters have these cycles of egg production at all times of the year. Other animals have a definite breeding season and it is only then that eggs are released from the ovary and it is only during this time that the female will mate with the male. Bitches, for example, have two breeding seasons a year, sometime in the spring and the autumn, when they are said to be on *heat* and bleeding from the uterus occurs. During this period of heat there is one cycle of egg production and this is the time when mating occurs. In between these times the bitch will not mate. The breeding season for roe deer is in July and August and is called the *rutting* season. There is one cycle of egg production and this is the only time that mating occurs. The time of the breeding season normally occurs so that the young are born at a favourable time of year, usually in the spring. Thus when deer mate in the autumn it means that the embryos will develop in the female and will be ready to be born in the spring. Bats mate in the autumn but their embryos only take a short time to develop in the female's body, and the sperms do not fertilize the eggs until early spring. Badgers mate in July or August in southern England and although the eggs are fertilized the embryos are not implanted in the uterus wall of the female until December or early January. The embryo then grows rapidly and the young are born eight weeks later.

Just what controls when the breeding season will occur is not easy to explain. The seasons and the length of day may have an effect although animals living where there is little change in the length of day will still have a definite breeding season. It is obviously important that when only one or two cycles of egg production occur during a year, there should be a good chance of the eggs being fertilized. A definite breeding season, when mating occurs and coincides with egg production, makes reproduction much more likely.

Introducing living things

Little mention has been made of humans and this is because they are different from the animals mentioned.

The menstrual cycle, about which there is more in section 5.52, is every 28 days. It will probably start between the ages of 10 and 13 years and occurs throughout the life of a woman until she is 45 to 50 years old.

But humans do not have a breeding season. Men and women have *intercourse*, as mating in humans is often called, not only at the time of egg production. Humans have an attraction and feelings for each other quite different from other animals and may have intercourse at any time.

5.43 How long does it take for a mammal to develop in the uterus?

Some female mammals, like human females, usually produce one egg at a time. That is why humans, elephants, cows, and seals usually give birth to a single young one while such mammals as dogs, cats, rabbits, mice, and rats produce more than one egg at a time and so give birth to several young.

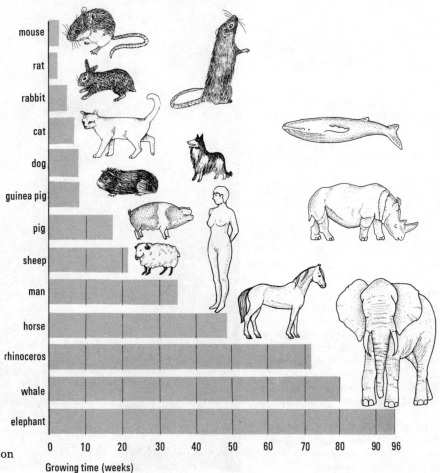

Figure 87
A chart showing the gestation periods of some mammals.

Growing time (weeks)

The length of time between fertilization of the egg in a mammal and birth is called the *gestation period* or *pregnancy*.

Q2   What factors would you suggest are related to the different lengths of gestation periods in mammals?

Pouched mammals such as kangaroos are unlike the mammals mentioned so far in that there is no placenta or only a very small one. Once the food store in the egg is used, the embryos emerge and continue their development in the pouch of the female. The gestation period in pouched mammals is, therefore, very short. It is 12 days in the Virginian opossum and only 40 days for the much larger grey kangaroo, where the new-born young are only about 2.5 cm long. (See *figure 88*.) The time spent in the pouch may be much longer than the gestation period.

Figure 88
A newly born kangaroo. As a very small embryo in a very early stage of development, it leaves the uterus and climbs through the thick hair of its mother's abdomen into her pouch. There it fastens itself on to a teat and gets a supply of milk for several months until it is fully grown.
Photograph, Division of Wildlife Research, Commonwealth Scientific and Industrial Research Organization, Canberra.

Introducing living things

Protection of the developing embryo

The needs of a developing embryo are food, water, air, warmth, and protection. We have seen that in the hen, the embryo chick develops inside the egg; this contains yolk on which the embryo feeds while it is growing, and white (albumen), which surrounds the embryo, providing water and protecting it against vibrations. The hen's egg also has a hard shell which protects the content of the egg, at the same time allowing air to pass into and out of the egg. Different kinds of birds lay different sizes of eggs. A large bird such as an ostrich lays a very large egg, while a small bird such as the wren lays a small egg. The smallest bird of all, a humming bird, lays an egg only about 12 mm long.

	Incubation time (days)	Length of fully grown bird (cm)
hawfinch	$9\frac{1}{2}$	18.0
cuckoo	$12\frac{1}{2}$	30.0
song thrush	13–14	21.5
humming bird	13–18	5.0–23.0
rook	16–18	48.0
teal	24	36.0
mute swan	36–40	144.0
ostrich	40–45	up to 230.0
Manx sheerwater	52–54	38.0
king penguin	56	93.0
fulmar	57	50.0
emu	58–61	200.0
kiwi	70–75	70.0

Table 9
A list of some birds whose incubation times are known and the lengths of the fully grown bird.

Is there a relationship between the size of the bird (and therefore the size of the egg) and the length of time the egg is incubated?

What other factors may affect the length of the incubation time?

Reptiles such as tortoises, and most kinds of snakes and lizards also lay eggs which contain yolk and a watery fluid, the egg being enclosed by a leathery or hard shell. In these animals the embryo inside may become well developed before the egg is laid. Amphibians such as frogs, newts, and salamanders lay their eggs in water like *Xenopus*. The egg has no shell, contains some yolk, and is surrounded by a layer of jelly. Most fish also lay eggs in water, the egg being protected by a jelly or leathery covering.

Figure 89
Trout eggs developing. The eggs
are nearly ready to hatch and the
embryos can be seen inside the
transparent eggs.
Photograph, W. T. Davidson/
Frank W. Lane.

To make a summary of the ways in which developing
embryos are provided for make a larger copy of *table 10* in
your notebook and complete it.

Animal group	Food	Water	Air	Warmth	Protection
Fish					
Amphibians					
Reptiles					
Birds	yolk	white	shell	body of parent	the white and the shell
Mammals					

Table 10
A table to contain a summary of
the ways in which developing
embryos are provided for.

5.5 How do human beings reproduce?

Human beings, like other mammals, give birth to their
young. In addition, human parents have to spend a much
longer time looking after their young than any other
animal. It takes a long time to learn to talk to and
understand other people and to be able to work with them.
Also a human child needs to learn a great deal of
information and skills such as walking, talking, reading,
and writing, before he is able to look after himself.

5.51 Am I normal for my age?

Having come near to the end of childhood, you have learned a great deal, and are gaining control of your body and behaviour. You now enter *adolescence* and you begin to grow very quickly. Suddenly you need to master a whole new set of skills and you have a whole new set of learning to do and a new range of feelings to cope with.

One of the things you will be thinking about a lot is how fast you are growing. We call this the 'growth spurt', which occurs at adolescence. Look around the people in the class – are they all the same size? No, of course they aren't. Some are tall, some short, and some in between. This is because people reach their adult height at different times and do so at different speeds. Some people in your class may have finished most of their growth, others may not yet have begun this sudden increase, others may be in the middle of it. But humans do not simply grow, they think and feel about how they are growing; some worry because they are bigger than everyone else and some because they are smaller, and many more wonder whether they are normal or not.

Who's normal?
Is it normal to have your growth spurt at eleven or at sixteen? The answer is neither or both. In human beings nothing is more abnormal than being average for everything – carry out the following investigations on yourself and put all the results together and plot them on class graphs. You can then see where you are in relation to the *norm* (the average results) of your class – but perhaps your class isn't normal when compared with the rest of the country!

Investigations	Result	Position on class graph		
		low	*middle*	*high*
1 Find your mass	kg			
2 Measure your height	cm			
3 Breathe in as far as you can, then measure right around your chest	cm			
4 Breathe out as far as you can, then measure right around your chest	cm			
5 Write down what size shoes you take				
6 Write down the time taken to do your last English homework				
7 Write down the time taken to do your last maths homework				
8 Dip 2 cm of drinking straw into the salt solution provided, then squeeze a few drops onto your tongue; decide whether it tastes good (G), or horrible (H), or has no taste (N)	G, N, or H			

Now plot your results on class graphs and answer the following questions.

Q1 Does the graph of mass show a peak?
If not (as in the results for one class, shown here in *figure 90*), can you think of reasons why?

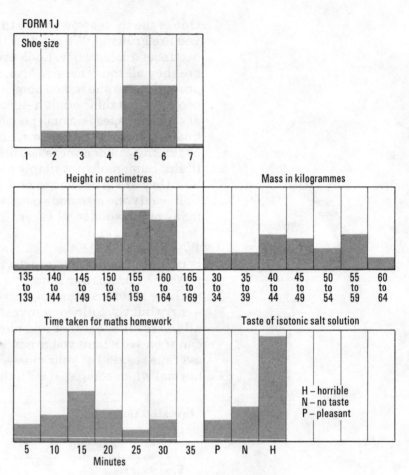

Figure 90
How one class held an investigation into human variation.
With thanks to the Lady Margaret School, Parson's Green, London SW6.

Q2 Why is everyone in the class not the same height?

Q3 What can you discover from the graphs of chest measurement?

Q4 In the school results shown in *figure 90*, there is a more obvious peak in shoe size than in height – can you give two reasons why this is so? Is your class graph the same?

Q5 Does the peak for maths homework include the same people as the peak for English homework? Are they the same people who were in the height peak and the shoe peak?

Q6 In tasting the salt solution *you* have to decide what is *abnormal*. Some people do not like the taste, some say it has no taste at all, others think it tastes nice. What do most people in your class think?

Q7 Make a list of other things in which human variation is normal; for example, some people are allergic to penicillin or to bee stings. Some never catch colds, others do all the time.

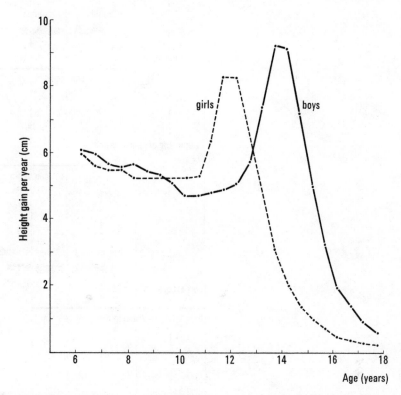

Figure 91
A survey carried out into the growth spurt in the height of girls and boys.
From Tanner, J. M. (1962) Growth at adolescence, 2nd edition, Blackwell Scientific Publications.

But the most important thing to remember is that humans, whether tall, fat, thin, or short, are people. Their appearances cover a wide variety of personalities. These personalities cannot be understood by weighing or measuring!

Look at *figure 91*, which shows the results of a survey of height increase. Then try to answer these questions.

Q8 Who began their growth spurt earlier, boys or girls?

Q9 Who finished their growth spurt first, boys or girls?

Q10 Who had the biggest gain in height in any one year, boys or girls?

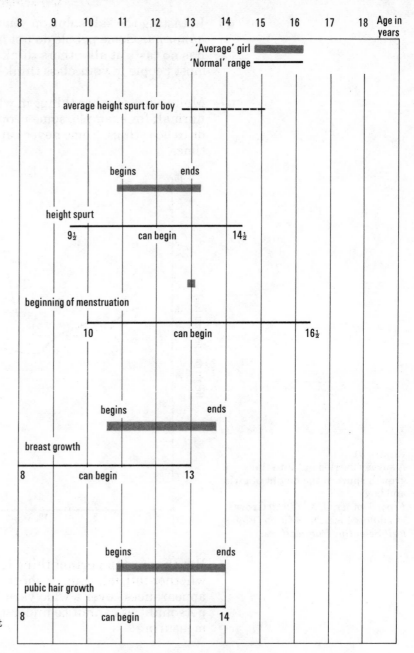

Figure 92
The 'normal' range of development
of an 'average' girl.

Same pattern, different rate
You can see that boys and girls both have the same pattern
of growth – their increases in height take place over about
the same number of years, and although boys grow most in
any one year, the difference is only about 2 cm. But if you
look at the people in your class, you may realize that their
rates of growth do not fit those in the graph. This shows
only the *average* times of growth spurts and very few human
beings are average. All human growth follows the same
pattern but each individual sets a growth rate of his own.

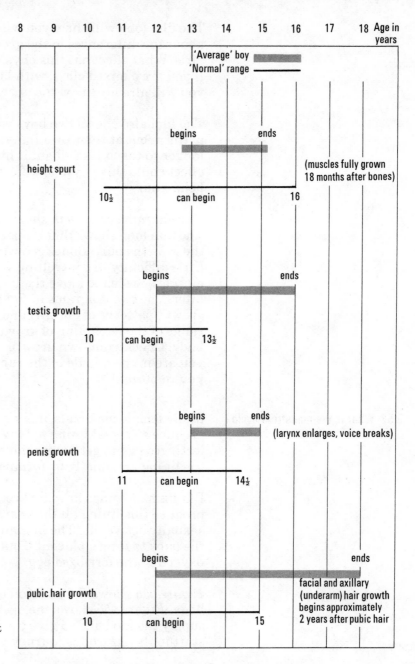

Figure 93
The 'normal' range of development of an 'average' boy.

The film loop 'Growth and development of humans' was made by taking pictures of some boys and girls at intervals of six months. The film makers tried to put them in exactly the same positions, but of course this is not possible when children are growing fast; for example, the boys' shoulders can be 3 cm higher between one photograph and the next. Also, all the photographs were printed on a ciné film and so you can observe humans growing up over thirteen years in only four minutes!

The film loop will show you what you have already learned about the differences in the growth spurt between boys and girls. What effect has this physical difference on behaviour? If you are a mixed class, with large girls and small boys, you will already know the answer to this question!

The film also shows two boys who develop in the same pattern, but at their own individual rates – one boy takes longer to reach his full adult height than the other. What effect could this physical difference have on their behaviour?

Normal range of growth spurt
The film loop shows that all parts of the body change during the growth spurt. Bones grow first and you might have a little difficulty in controlling your movements as the larger muscles needed to move the larger bones only grow much later. The two diagrams in *figures 92* and *93* show the growth patterns of one girl and one boy and also the normal range of the beginning of growth in various parts of the body. Check your own growth – it doesn't matter whether you are in the middle of the range or at one of its ends – you're normal!

5.52 What is the menstrual cycle?

Many female mammals, during their breeding season, produce a special lining for the uterus. This will help the fertilized egg to get nourishment, first by burying itself in the lining and finally by forming a placenta.

But human beings have no breeding season, and women produce this lining all the year round, during all their sexually active life. The human uterus lining not only helps the baby to form a placenta, but also supplies both food and oxygen to the fertilized egg.

Figure 95a shows the position of the uterus in the human body. *Figure 95b* shows the openings of the glands which supply the food and so give the very new human a good start in its hazardous journey to birth.

The process of menstruation
Food for humans is also food for bacteria, and this is, perhaps, one reason why it is a good thing that the human uterus lining is replaced every 28 days (or every 21 to 35 days, since this is the normal range of human variation).

Q11 Can you think of other advantages that result from the lining being renewed so often?

Remember that all mammal eggs are very small and very

Figure 94
A 12½-day-old human embryo.
*Photograph, Dr Landrum B.
Shettles.*

delicate, and the human egg needs to burrow into the
uterus wall.

Q12 How does the uterus lining grow in time for the fertilized
egg to be able to use it?

A very complicated set of biological mechanisms, including
body chemicals, blood vessels, and nerves, take part in an
almost unbelievable combined operation every 28 days.

In the diagrams in *figures 95b* to *95g*, you see the stages in
the cycle and the results of the working together of a large
number of biological processes.

It takes the body 5 days (3 to 7 days is the normal range of
human variation) to get rid of the old lining. About 9 days
after this the egg is fully grown, bursts out of the ovary, and
is usually caught by the egg tube. The uterus lining is fully
grown after about 26 days. After this, the blood vessels of
the inner layer of this lining shut off and the outer layer
begins to break up. These events are summarized in *figure
96.*

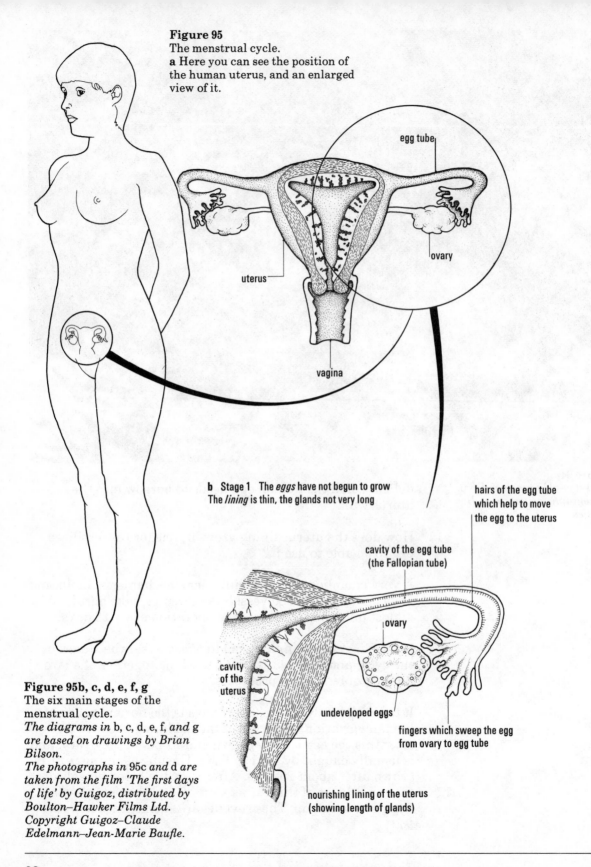

Figure 95
The menstrual cycle.
a Here you can see the position of
the human uterus, and an enlarged
view of it.

egg tube

uterus

ovary

vagina

b Stage 1 The *eggs* have not begun to grow
The *lining* is thin, the glands not very long

hairs of the egg tube
which help to move
the egg to the uterus

cavity of the egg tube
(the Fallopian tube)

ovary

cavity
of the
uterus

undeveloped eggs

fingers which sweep the egg
from ovary to egg tube

nourishing lining of the uterus
(showing length of glands)

Figure 95b, c, d, e, f, g
The six main stages of the
menstrual cycle.
The diagrams in b, c, d, e, f, *and* g
*are based on drawings by Brian
Bilson.*
The photographs in 95c *and* d *are
taken from the film 'The first days
of life' by Guigoz, distributed by
Boulton–Hawker Films Ltd.
Copyright Guigoz–Claude
Edelmann–Jean-Marie Baufle.*

Introducing living things

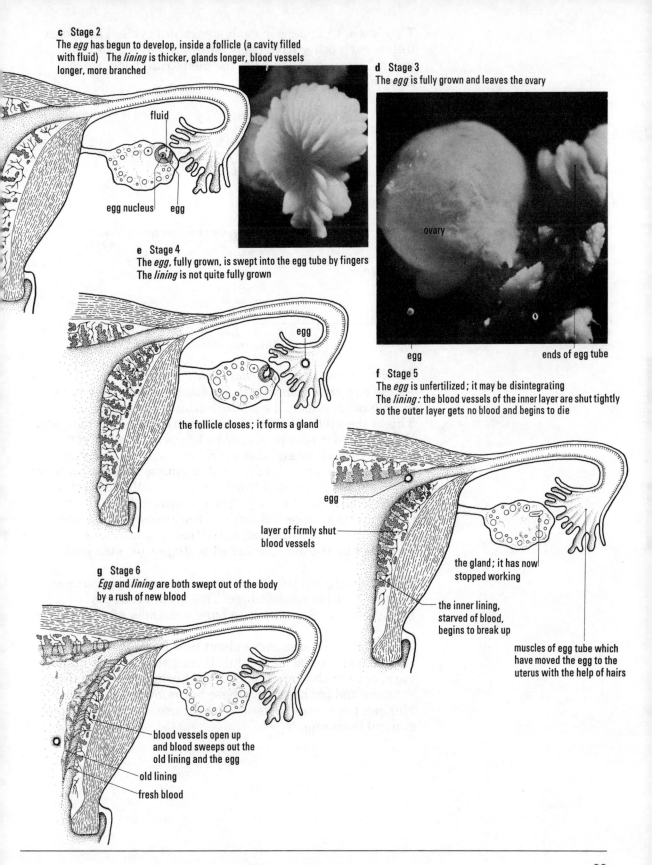

c Stage 2
The *egg* has begun to develop, inside a follicle (a cavity filled with fluid)   The *lining* is thicker, glands longer, blood vessels longer, more branched

fluid

egg nucleus   egg

d Stage 3
The *egg* is fully grown and leaves the ovary

ovary

egg   ends of egg tube

e Stage 4
The *egg*, fully grown, is swept into the egg tube by fingers
The *lining* is not quite fully grown

egg

the follicle closes; it forms a gland

f Stage 5
The *egg* is unfertilized; it may be disintegrating
The *lining*: the blood vessels of the inner layer are shut tightly so the outer layer gets no blood and begins to die

egg

layer of firmly shut
blood vessels

the gland; it has now
stopped working

the inner lining,
starved of blood,
begins to break up

muscles of egg tube which
have moved the egg to the
uterus with the help of hairs

g Stage 6
Egg and *lining* are both swept out of the body by a rush of new blood

blood vessels open up
and blood sweeps out the
old lining and the egg

old lining

fresh blood

How living things develop

This is what happens in each menstrual cycle. The cycles follow each other from early adolescence to middle age; the normal age range at which the menstrual cycle stops is from 40 to 55.

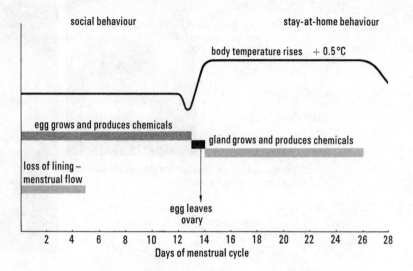

Figure 96
A diagram of events in the menstrual cycle.

Hormones and the menstrual cycle
As the egg grows, the ovary produces a chemical which, among other effects, helps the lining of the uterus to grow. This is the kind of body chemical known as a *hormone*, and this particular one, produced by the ovary, can affect human behaviour and make girls and women feel much more sociable and friendly – but of course it doesn't always do this! This is because this hormone is only one factor which influences human behaviour, and it does not always overcome the major influence of the brain, which decides the pattern of most human activities. The hormone produced by the ovary is called *oestrogen* (ee-stro-jen).

When the egg has left the ovary, another kind of hormone is produced called *progesterone*. This often produces behaviour which improves a baby's chances of survival.

Later, you will learn more about how hormones affect women's and girls' behaviour. But remember that humans vary, both in the amounts of these hormones which they produce and in the effects these hormones have on them. Nor can hormones be used as an excuse – usually you can control their effects.

Temperature changes in the menstrual cycle

You have seen from *figure 96* that temperature rises during the latter part of the cycle very slightly. How can this affect behaviour? Again you must remember that human beings vary and not all girls will be affected in the same way by this rise. There is evidence that some feel much warmer and much more sleepy in the mornings; girls might need an alarm clock more at this stage in the cycle if they are not to be late for school! At this time, sweat glands work more actively and girls need more baths rather than less. Some old wives' tales would have us believe that girls should not bathe during menstruation!

Why can't you tell when girls are menstruating?

Human beings vary so much that all girls react differently to the hormones they produce and every girl produces different amounts of them. Some girls look more beautiful during menstruation – their skin glows, their hair looks just right, and they feel good. Others get very spotty during this time, their hair is greasy, and they feel generally fed up. Lots of girls feel fed up and are spotty at many other times in the cycle!

5.53 How fertilization takes place

When girls first menstruate, they may feel that it is a nuisance. After all, they will only use a very small number of the eggs and linings which they will produce during their lives. But at least they know approximately when menstruation is going to happen, whereas when boys first begin to produce sperms, they do not do so in regular cycles. During a man's lifetime, millions and millions of sperms will be produced, of which only a few will ever fertilize eggs.

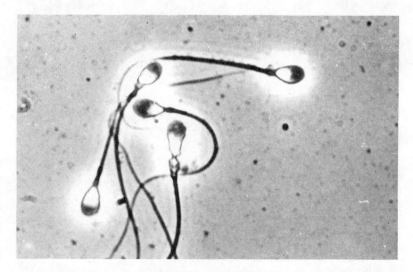

Figure 97
Human spermatozoa.
Photograph, Professor J. L. Hancock.

Human sperms are produced in the testes and, when fully grown, lie there in a fluid which gives them food and oxygen. Millions of sperms are produced, of which only one is needed to fertilize an egg. When boys first begin to produce sperms, they do not realize it until the sperms leave the overcrowded testes and come out through the penis. They are accompanied by the nourishing fluid and by hormones which make the sperms active. These hormones come from the glands shown in *figure 100* (page 104).

But how can the sperms reach the egg?
Most of the time, the penis is limp, but when sperms are about to leave the body, the blood vessels inside the penis become wider, and as they fill with blood, it rises and becomes stiff. It can then be put inside the vagina, thus forming a tunnel for the passage of the sperms to the egg. The tube which releases urine from the bladder closes when sperms are travelling to and through the penis.

The passing of sperms into the vagina has a variety of names. For example, it can be called making love, coitus (co-ittus), copulation (see page 81), or sexual intercourse.

Leonardo da Vinci made perhaps the first anatomical drawing of human sexual intercourse. You can see this in *figure 98*.

Figure 98 *(opposite)*
Leonardo da Vinci's drawing of coitus. (There is one anatomical detail which is not correct, and that is the channel which he portrays, joining the woman's breast to her uterus. There is no direct channel, but in the fifteenth century, this was thought to exist.)
Reproduced by gracious permission of Her Majesty the Queen.

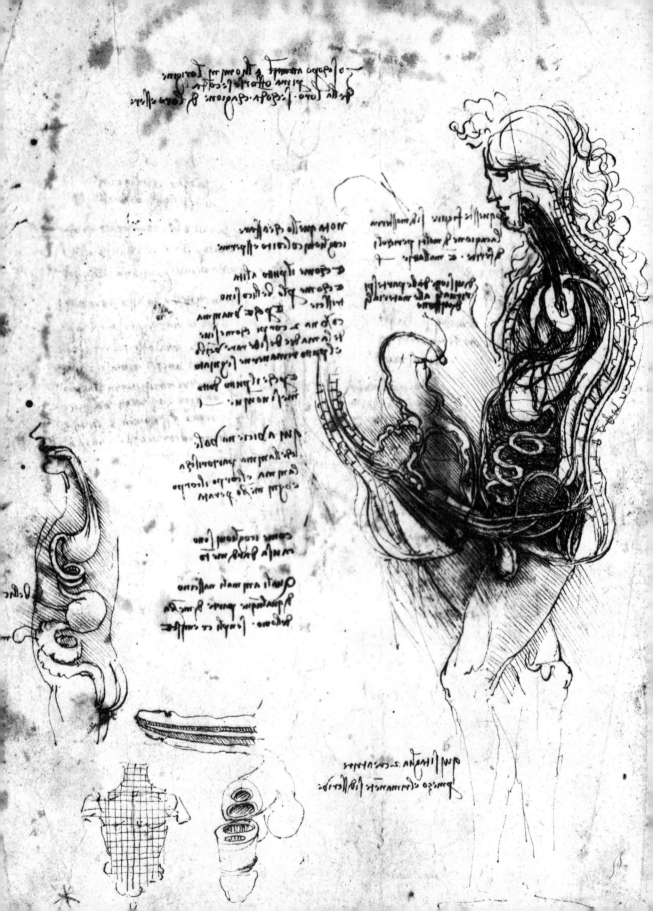

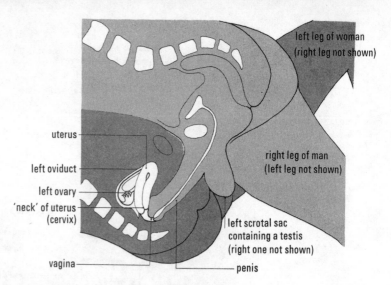

uterus

left oviduct

left ovary

'neck' of uterus (cervix)

vagina

left leg of woman (right leg not shown)

right leg of man (left leg not shown)

left scrotal sac containing a testis (right one not shown)

penis

Figure 99
A diagram to show the positions of the organs during sexual intercourse.

Figure 100
The male reproductive organs. *Based on Hancock, J. L. 'The sperm cell', originally published in* Science journal, *June, 1970.*

Boys often have no warning of when their penis is likely to rise and become erect and stiff. They soon learn that all sorts of things, mostly things that they see, can make this happen. If such things happen to you it does not mean that you are in any way peculiar; they are due to a new piece of body machinery which is not fully working yet and also, of course, to human variation.

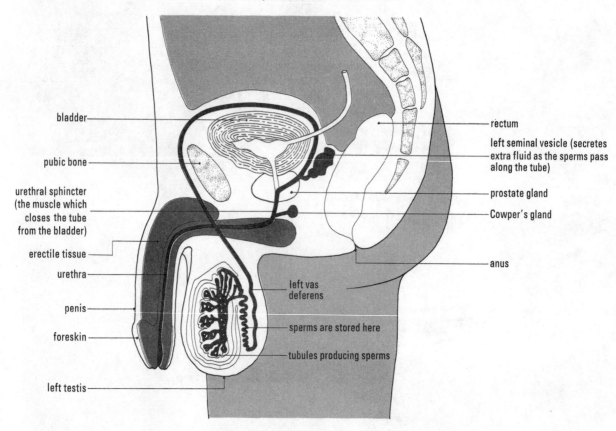

bladder

pubic bone

urethral sphincter (the muscle which closes the tube from the bladder)

erectile tissue

urethra

penis

foreskin

left testis

left vas deferens

sperms are stored here

tubules producing sperms

rectum

left seminal vesicle (secretes extra fluid as the sperms pass along the tube)

prostate gland

Cowper's gland

anus

Introducing living things

Boys produce their first sperms at different ages, and they all produce different amounts at different times. Sometimes the sperms leave the body when the penis is rubbed by the hand. At other times the sperms may come out while a boy is asleep at night, accompanied perhaps by romantic dreams. This is in no way harmful, although it was once thought that boys should be ashamed of such dreams.

On the other hand, many boys may wake with an erection of the penis and not know why. It may be just because they want to urinate.

II Male hormones and behaviour

At the same time that it produces sperms, the human testis also produces small quantities of powerful hormones which affect males physically and also affect behaviour. You have already met those produced by human females, oestrogen and progesterone; the human male produces hormones called *androgens*. When boys begin to produce androgens, their vocal cords become longer and their larynxes bigger, so that their voices become deeper. Some boys might find it difficult to control their voices when this change is happening.

Androgens also increase the activity of the small sebaceous glands in the skin which produce a greasy substance called sebum. Boys may find that they have more blackheads, acne, and spots; acne can leave the skin pitted.

Androgens affect other animals as well as humans. When the testes are removed from bulls and stallions, for example, the animals become much less aggressive and do not want to mate.

You may have noticed that boys are usually more boisterous, active, and aggressive than girls. While androgens certainly play a part in causing this behaviour, the brain and the opinions of the people humans live with also play a part in deciding the behaviour of human males.

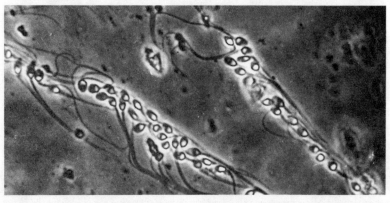

Figure 101
Human sperms swimming (× 1500).
Photograph from Parish, W. E. and Ward, A. (1968) Journal of obstetrics and gynaecology of the British Commonwealth, *75, 1089.*

III Which sperms survive?

Although the sperms are protected by the penis, which lets them pass safely from one safe environment to another, relatively few ever reach the egg. Some are not formed properly and others cannot use the energy in the surrounding nourishing fluid to convert it into their own movement. Those which can, and which can swim, by lashing their tails, up the vagina, meet a warmer temperature than they were used to in the testes: the testes hang outside the body and are cooler than the rest of it. Also, the fluid in the vagina is different from that in which the sperms were swimming; for instance, it may be more acid. If so, and if some of the sperms come in contact with the vaginal acid, they are made inactive by it.

Having got through the vagina, there is still a long way for the sperms to swim. A sperm usually fertilizes the egg right at the very end of the Fallopian tube, where there are finger-like projections which sweep the egg into the egg tube. *Figure 102* compares the sizes of the egg and sperm.

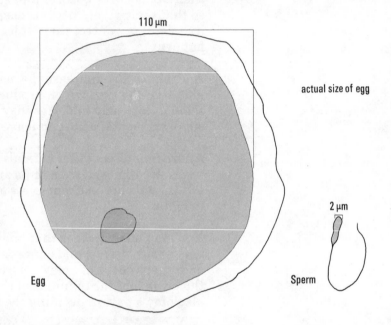

Figure 102
A comparison of the sizes of a human egg and a sperm.

Introducing living things

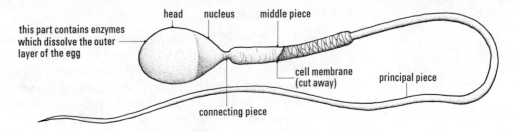

head   nucleus   middle piece

this part contains enzymes
which dissolve the outer
layer of the egg

cell membrane
(cut away)

principal piece

connecting piece

Figure 103
A detailed diagram of a human
sperm.
*Based on Hancock, J. L. 'The
sperm cell', originally published in*
Science journal, *June, 1970.*

IV Egg meets sperm

Several sperms reach the egg, but only one fertilizes it. That
is, the nucleus of the sperm joins with the nucleus of the
egg. *Figure 104* shows the two nuclei just before they join
together to form one cell.

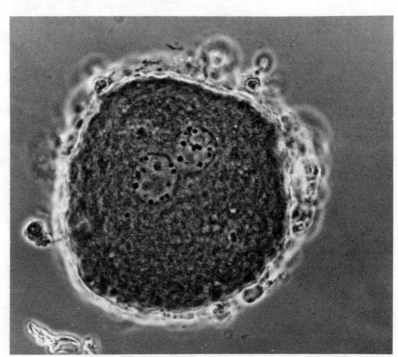

Figure 104
A recently fertilized human egg.
It is about 24 hours old. You can
still see the two pronuclei,
one belonging to the egg and one
from the sperm. When they fuse
they form a nucleus. This fertilized
egg would normally divide within
about 48 hours.
Photograph, Dr Z. Dickmann.
From Dickmann, Z. et al. (1965)
Anat. rec. *152, 293.*

This one cell divides into two and a baby begins its
growth. From this one cell will eventually come the
millions which make up the human baby, and each nucleus
will contain a set of patterns, some from the mother and
some from the father, which will decide what the baby
will look like. It is these patterns which are stored in the
nucleus of the egg and the sperm.

V Sexual emotions

The effect of new hormones in our blood streams at adolescence changes our emotions and moods. Jealousy, anger, love, and hate, for example, may all seem much stronger now than they did before. Boys and girls begin to be attracted to each other, and loving relationships develop, some of which may deepen and last.

In a lasting and loving relationship, sexual feelings and sexual intercourse are very important. They give pleasure both to the woman, and to the man. They are concerned not only with making babies, but with living together, caring for each other, and sharing things, as part of a human family.

Sexual growth is not completed until the late teens (the normal range of variation is 17 to 23) and while people under that age can and do have both sexual intercourse and babies, many are disappointed because they are not satisfied by either event. These experiences need not only a mature body for full enjoyment but also a mature mind and personality.

Figure 105
From one cell to a ball of cells.
From the film 'The first days of life' by Guigoz, distributed by Boulton–Hawker Films Ltd. Copyright Guigoz–Claude Edelmann–Jean-Marie Baufle.

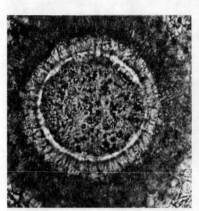

a At the time of ovulation

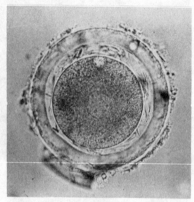

b The 12th to 24th hours

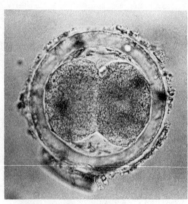

c The 30th hour

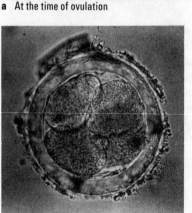

d The 2nd day

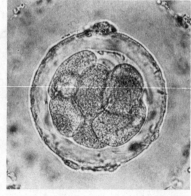

e After 2½ days

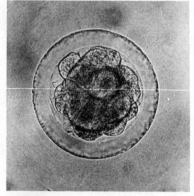

f The 4th day

Introducing living things

Every human egg, when fertilized, follows the same pattern of development but, of course, each individual fertilized egg develops at a different rate.

I Organization

Basically what happens is that the fertilized egg divides as it rolls down the egg tube to the uterus. You can imitate the process by rolling a piece of Plasticine about the size of a toffee into a ball. Then divide it into two balls, then each of these into two, and so on until you have a little ball of little balls! This is the stage which the fertilized egg has reached when it gets to the uterus. You may be able to see the film 'The first days of life' which shows actual pictures of a fertilized human egg doing this.

Twins and multiple births
Sometimes when the egg has reached the two-cell stage, **both** cells develop into embryos. These both have the same share of the nucleus of the egg and the sperm and will become *identical* twins. They must be of the same sex.

Sometimes the ovary releases two eggs at once. These are fertilized by different sperms and become *non-identical* twins. They can be boy and girl twins.

Q13 Can you now think of two ways in which the quintuplets shown in *figure 106* might have been begun?

Figure 106
Quintuplets.
Photograph, Syndication International.

From fertilized egg to embryo

The cells in the little ball, all of which look the same, turn into a digestive system, a blood system, a nervous system, bones, skin, muscles, all in the right order and in the right place. Scientists have only very little idea how it is done. It is a very complicated process indeed, and it is remarkable how successful it is.

The drawings in figures 107 to 110 are based on Hamilton, W. J., Boyd, J. D., Mossman, H. W. (1966) Human embryology, Heffer.

The ball of cells hollows out

As you can see from *figure 107*, the ball of cells hollows out, and after a great deal of reorganization, at 28 days old it looks like the picture in *figure 108*.

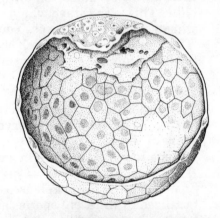

Figure 107
The ball of cells hollows out . . .

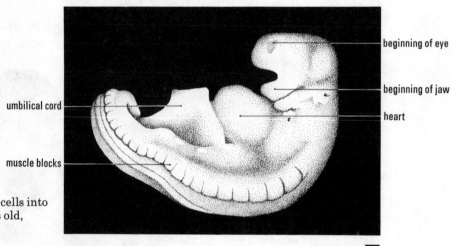

beginning of eye

beginning of jaw

heart

umbilical cord

muscle blocks

Figure 108
. . . and begins to organize cells into systems. Here it is 28 days old, 3.4 mm long.

actual size

Introducing living things

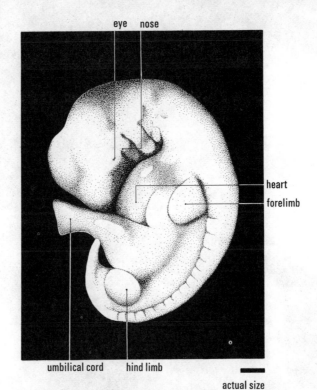

eye nose

heart

forelimb

Figure 109
34 days old, 6.7 mm long.

umbilical cord   hind limb

actual size

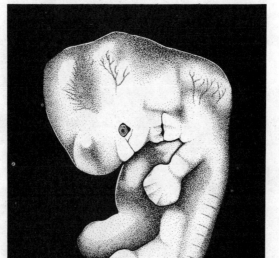

you can label this
yourself now

Figure 110
40 days old, 13.4 mm long.

actual size

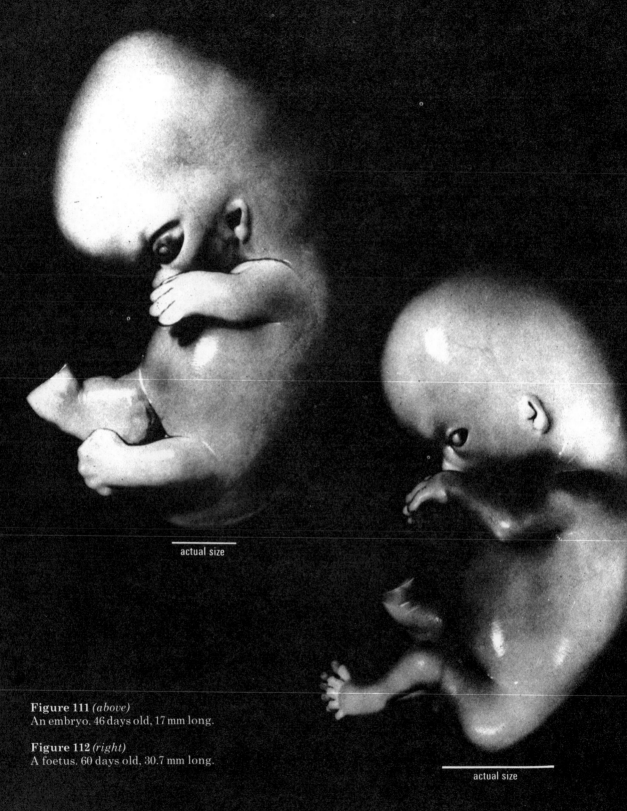

Figure 111 *(above)*
An embryo. 46 days old, 17 mm long.

Figure 112 *(right)*
A foetus. 60 days old, 30.7 mm long.

actual size

actual size

Q14   What is the difference between an embryo and a foetus?

Figures 111 to 114 are reproduced by courtesy of Professor W. J. Hamilton, J. D. Boyd, H. W. Mossman, and W. Heffer & Sons.

II Growth

By the time the unborn human baby is about three months old, it has finished most of its organization, as you can see from the 44 mm foetus in *figure 113*. Now it really begins a growth spurt.

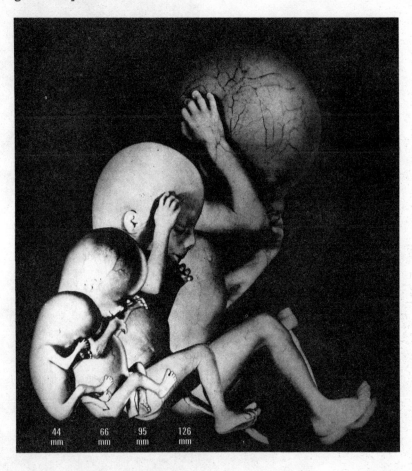

44 mm   66 mm   95 mm   126 mm

Figure 113
Human foetuses from 3 to 5 months old.

III The baby's life-support system

Look back to section 5.52, 'What is the menstrual cycle?', and remind yourself how the baby gets food and oxygen in the first few days of its life. It does not get very much, so it does not grow very much at this stage.

What the embryo must do after attaching itself to the uterus wall is to form many finger-like projections. These then become bathed in the rich blood supply coming from its mother and the placenta is formed.

Q15 What differences are there in the life-support systems of the two kinds of twins shown in *figure 116*?

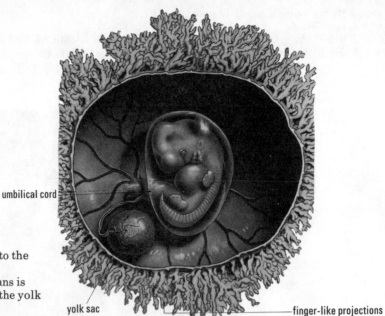

Figure 114
How the embryo attaches to the uterus.
Note: the yolk sac in humans is not put to the same use as the yolk of a hen's egg.

umbilical cord

yolk sac

finger-like projections

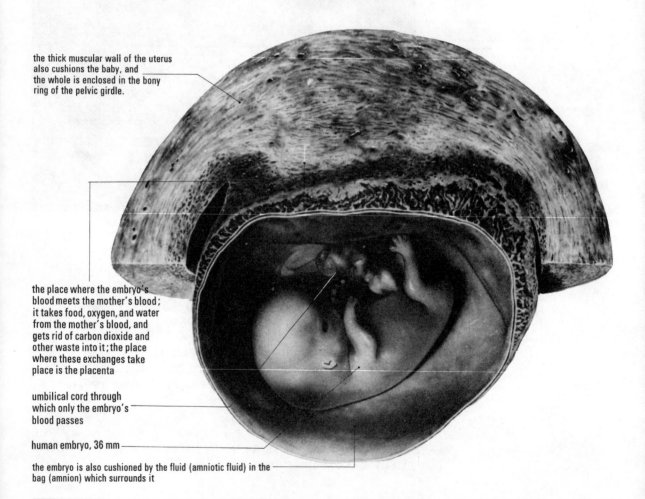

the thick muscular wall of the uterus also cushions the baby, and the whole is enclosed in the bony ring of the pelvic girdle.

the place where the embryo's blood meets the mother's blood; it takes food, oxygen, and water from the mother's blood, and gets rid of carbon dioxide and other waste into it; the place where these exchanges take place is the placenta

umbilical cord through which only the embryo's blood passes

human embryo, 36 mm

the embryo is also cushioned by the fluid (amniotic fluid) in the bag (amnion) which surrounds it

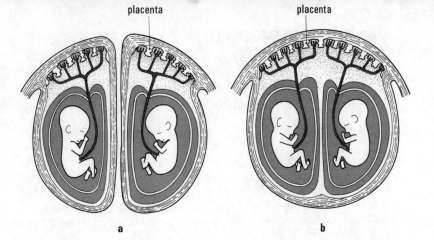

placenta   placenta

a   b

Figure 116
Twins in the uterus.
a Non-identical.
b Identical.

IV What pregnancy feels like

Earlier, you thought about how human beings vary and
how they have emotions. These two basic principles affect
pregnancy too. The hormone progesterone you learned
about in the work on menstruation is produced in large
quantities during pregnancy and, of course, affects
behaviour as well as helping to form the placenta and to
develop the breasts so that milk will be made – the baby's
first food. Another hormone is made by the placenta. This is
lost in the urine, and so helps doctors to know whether
pregnancy has begun or not.

Some women don't enjoy being pregnant. They may feel
sick before breakfast. Others are serene and happy, even
when they look very large, with the baby and placenta and
amniotic fluid inside them.

Mothers need to have careful medical checks during
pregnancy to make sure that all is going well. In particular,
doctors listen to the baby's heart through a special
instrument. Just before birth they see if the baby is lying
in the uterus in such a way that it can come out head first.
Can you see what might happen if it comes out feet first?
The X-ray of twins in *figure 118* shows that one of them will
probably come out feet first.

Figure 115
How a one-month-old embryo is
nourished and protected.
*Photograph, W. J. Hamilton and
J. D. Boyd, by courtesy of the*
Journal of anatomy, *London.*

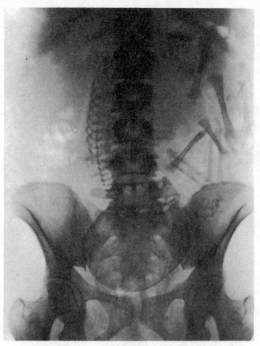

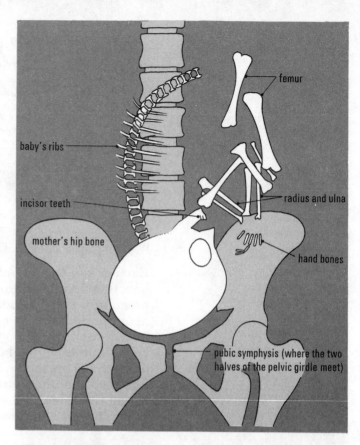

Figure 117
An X-ray and diagram of a single foetus.

Labels on diagram (Figure 117):
- femur
- baby's ribs
- incisor teeth
- radius and ulna
- mother's hip bone
- hand bones
- pubic symphysis (where the two halves of the pelvic girdle meet)

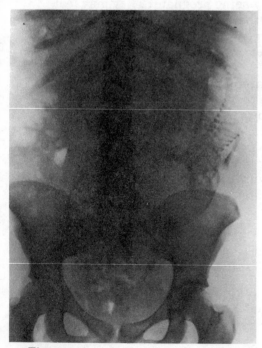

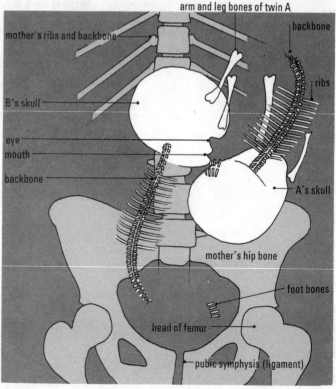

Figure 118
An X-ray and diagram of the foetuses of twins.

Labels on diagram (Figure 118):
- arm and leg bones of twin A
- mother's ribs and backbone
- backbone
- B's skull
- ribs
- eye
- mouth
- backbone
- A's skull
- mother's hip bone
- foot bones
- head of femur
- pubic symphysis (ligament)

5.55 The baby is born

When human babies are born, much hard work (labour) is involved for the mother. But the advanced stage of development of the human baby probably gives it a better chance of survival. It seems to stay in the uterus as long as it can, only leaving just before it gets too big to get out.

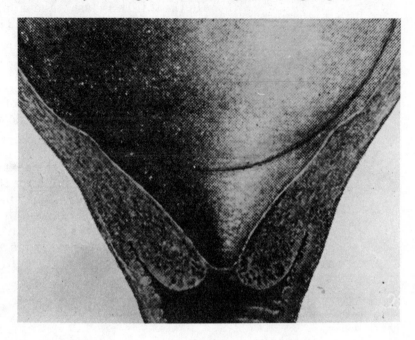

Figure 119
The first stage of labour, showing the neck of the uterus widening.
Photograph from the filmstrip 'Mor medbarn', Scandinavska Testforlaget, Stockholm.

I The stages of labour

The muscles along the whole length of the uterus shorten (contract) and so, eventually, push the baby out. But before this can happen the neck of the uterus (the *cervix*) must expand. These contractions of the uterus may be felt as a dull ache (labour 'pains') and are one of the signs that birth is about to begin. Sometimes the amnion breaks first under the pressure from these contractions and the amniotic fluid is lost.

During this first stage of labour when the cervix is expanding, the mother usually relaxes and lets her body do the work. There is little she can do to help. Hospitals and clinics hold special classes to teach mothers how to relax in the first stage of labour and also how to breathe in a way which should ease the second stage of labour when the mother can help the muscles of the uterus to push the baby out.

As the baby moves down the vagina, the difficult squeeze between the bones of the pelvic girdle begins.

When the baby is first born it may be blue, because it has not yet begun to breathe for itself and has stopped taking in oxygen through the umbilical cord. As soon as it takes its first breath, the blood takes up more oxygen again and the baby turns pink. Sometimes the nurse or doctor has to suck out the amniotic fluid from the nose and mouth with a tube, before the baby can take its first breath.

In the third and last stage of labour the placenta separates itself from the uterus wall and comes away.

II The effect of birth on the family

It is impossible to describe the feelings of human fathers and mothers about the birth of their babies but on the whole most of you can expect to enjoy the process like the parents in *figure 120*.

But for most mothers, the next year will mean a great deal of hard work in looking after the baby, and for the other children in the family it may mean rather less of their mother's care and attention than they had before he was born. Older children may think that the new baby is a noisy nuisance who keeps everyone awake at nights, but many, both boys and girls, really enjoy watching the new life grow

Figure 120
What are the feelings of the father?
The mother?
From the film 'Preparing for Sarah', Eothen Films (International) Ltd.

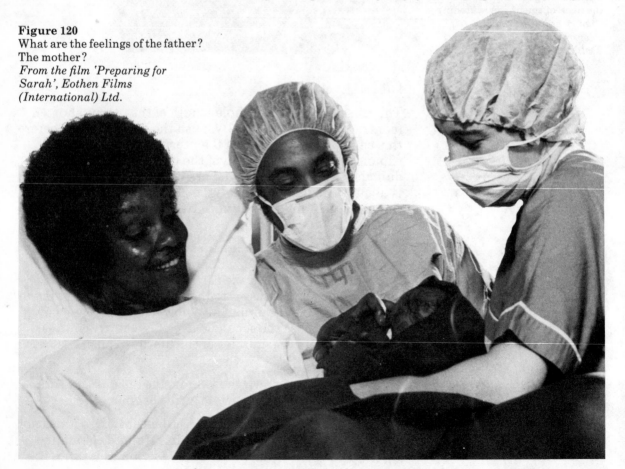

Introducing living things

up. And of course grandparents enjoy a new baby especially, because they have all the fun of playing with him and spoiling him without any of the hard work and sleepless nights.

Background reading

How parents look after their children

Some animals leave their young alone to fend for themselves and others look after them in varying ways. Sometimes it is the mother who looks after the young, sometimes the father, and sometimes both. Parent birds often share the work of nest building and feeding their young, and in fishes, the Siamese fighting fish male makes a bubble raft for his young and puts them back on it if they drop off, while the male seahorse (a bony fish) carries the young about in a pouch.

Figure 121 *(right)*
Male and female Siamese fighting fish and their bubble raft.
Photograph, Jeremy McCabe/ Frank W. Lane.

Kangaroos carry parental care to uncomfortable lengths. The young kangaroo hitches a ride even when he is almost too big to get into the pouch!

Human babies are, of course, very helpless when they are born. They need feeding and, in particular, to be kept warm as they cannot control their temperature as well as adults. Many books have been written on how to look after babies. At first they cannot digest solid food and have to have it all in liquid form, but as the muscles of their digestive system grow, enzymes are produced, and finally their teeth are able to chew food, so that by the end of the first year of life they can eat meals similar to those of adults.

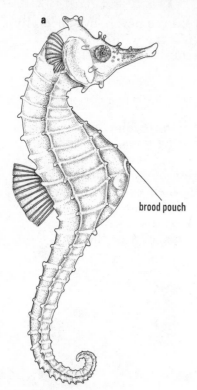

Figure 122
a A male seahorse.
b Young seahorses.
Photographs, Mondiale.

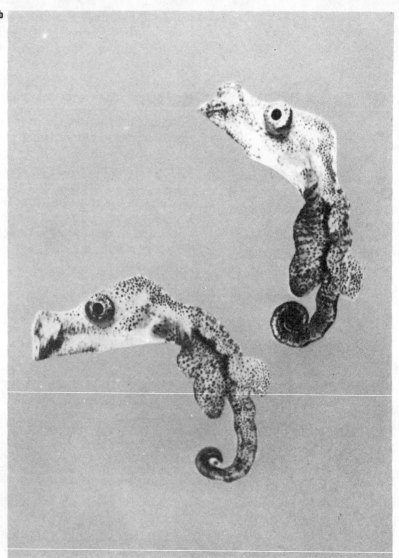

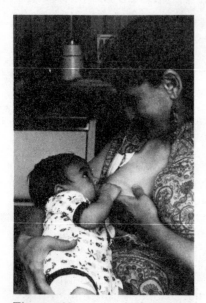

Figure 123
Photograph, Peter Weiss, Director,
ILEA Media Resources Centre.

But more than physical care, a human baby needs the kind of care that will help him to grow emotionally as well and to be able to communicate with other humans. How can he learn this, when he doesn't even begin to talk until he is about eighteen months old?

At first the baby cannot see things the right way up. Watch new babies trying to grab things, they usually miss! But when he has got his vision under control, the most important thing to him is to be able to watch another human face.

You can see in *figure 123* that the mother is smiling back at the baby, and that she is holding him close to her. This gives the baby the feeling that he is loved and protected.

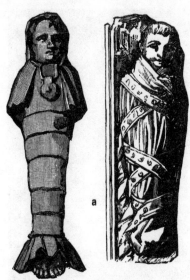

Figure 124
a How babies were swaddled in the past. *Left*, ancient Rome. *Right*, fifteenth-century France.
Photographs, from prints in the Mary Evans Picture Library.
b A modern method of wrapping a baby.
Photograph, Raymond Leng.

When does the baby begin to smile back? Well, often the first smile is due to a burp which can't quite manage to escape! But somewhere about the age of two or three months, even blind children smile, although they can see no one to imitate. Smiling babies make happy mothers, and happy mothers are more likely to do what the baby wants, but of course the baby soon learns to communicate by crying too!

When air first enters a baby's lungs he cries, not because he is miserable, but largely by unconscious action. Some babies seem to cry non-stop for the first few months, without any apparent reason. Some need to feel something firm around them, as the walls of the uterus once were, and like a parent's arms are (but obviously the mother or father cannot hold the baby *all* the time he needs to sleep). In Biblical times, babies were wrapped tightly in swaddling bands, and today, mothers wrap them firmly in a piece of sheet or blanket. Babies learn very quickly that if they cry, some parents will pick them up, and since a nice warm parent is better than a blanket any way, such parents do not get much sleep! (It is better to be a little hard hearted, use a wrapper of some sort, and be a lively loving parent the next day.)

b

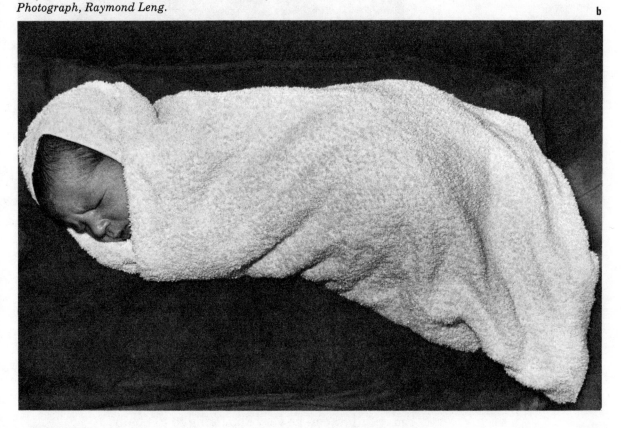

Figure 125
Children need both men and women to talk to and share things with. Here, Susan is showing her mother and father her playgroup. *Photograph, Pre-school Playgroups Association.*

Full parental care of human babies, then, means that they need to be physically cared for but also loved, cuddled, tickled, given moving things to look at, and talked to. Even when they cannot understand a word the parent says, it is the tone of voice which makes them feel secure and happy. Watch babies and their parents closely and see how they react to all these things.

Just as it is no good expecting a baby to walk until his muscles and bones are of the right level of development, it is no use expecting him not to use nappies until his bladder and rectum muscles can control the loss of urine and faeces. Human variation operates of course. Some babies are out of nappies quicker than others, but it is really not worth bothering about until the baby is two years old. It is not surprising that some mothers lose patience with their babies – and the breakdown of living communication which may follow can be very damaging to the baby.

Children learn more from their parents than they do at school – and human parents need to know just how much their children can understand at various ages so that they do not confuse them. Some schools now send their older pupils to work with experts in play groups and nursery groups. There they see how to help babies to learn through playing with sand, water, paints, and toys. Not only girls do this but boys as well. Once it was thought not quite manly for fathers to have much to do with their young babies. The fathers used to wait until they could go fishing, play football, or play with train sets before they took an interest in them. More and more fathers today (perhaps because more see their babies born?) take part in the process of helping babies to learn and communicate. They also help to feed them and wash them.

The next stage in human parental care is helping the child to communicate with people outside his family. You may have noticed that when very young babies are playing together they show little regard for each other, bash each other, take toys away, and generally cause havoc until their mothers calm them down. If you are still behaving like this you really will be abnormal! This social learning is essential for humans, and nursery school and play groups also help here, as does school later in life.

Figure 126
These children have the confidence to venture from the security of home to explore and enjoy the world about them. It will be a long time before they are independent but they have begun.
Photograph, Malcolm Shifrin.

The central aim of human parental care is to help the child to become independent in every way and, at the same time, able to work with and live with a variety of people, without which no human can survive. One of the hardest things for a mother to do is to realize that her child needs her no longer and is old enough to run his own life, and this is usually just as hard for fathers too. But this is just as essential a part of parental care as baby foods and teddy bears.

How plants multiply

The two previous chapters have been concerned with the ways in which animals reproduce. In this chapter you are going to look at the ways in which plants multiply.

6.1 How do new plants grow from a parent plant?

1 Take a single plant of duckweed and make a drawing of its outline, dating your sketch.
2 Transfer the plant you have drawn to the culture solution provided, using a brush so as not to damage it. The culture solution will provide the food substances necessary for the duckweed to grow.
3 Leave the jar in good light on a window sill for a week and then draw the plant again.

Q1 How many new 'leaves' have developed?

Q2 How many of these 'leaves' have grown a root?

Detach one of the 'leaves'. Keep it in the same jar, in the light, for at least a week. Examine it again.

Asexual reproduction is the production of an independent offspring from one individual only.

Q3 Can duckweed reproduce asexually?

Now study the drawing of *Bryophyllum daigremontianum* in *figure 128*, or better still, a living plant, if one is available.

Q4 Think of the definition of asexual reproduction. Is it taking place here?

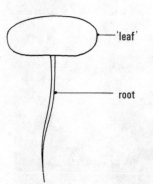

Figure 127
A single plant of duckweed.
(*Lemna minor*).

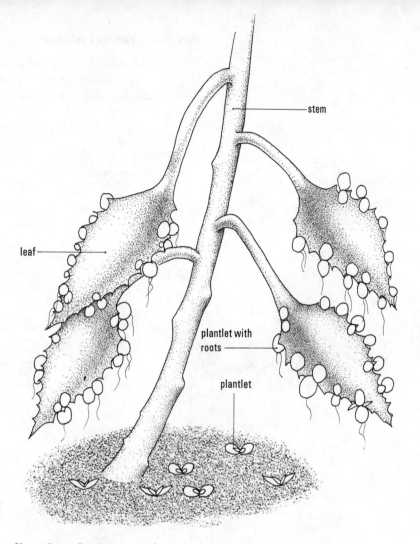

stem

leaf

plantlet with roots

plantlet

Figure 128
Bryophyllum daigremontianum.
Several plantlets have dropped off the parent plant and are growing in the soil below.

New plants from stem cuttings

A sprig of willow or 'wandering sailor' in a vase in the house will often grow new roots very quickly. Devise and set up a simple experiment to compare how quickly roots will form on a range of different stem cuttings.

Q5   What special precautions would you take when designing the experiment so that the results could be compared?

Q6   Would you expect to get the same results at all times of the year?

There are special chemicals (plant hormones) available which encourage cuttings to root. Gardeners use these and you may be given the opportunity to experiment with them.

New plants from root cuttings

You could also try to grow new plants from pieces of root.
1 Dig up a dandelion carefully so that you have the whole
   plant with its root complete.

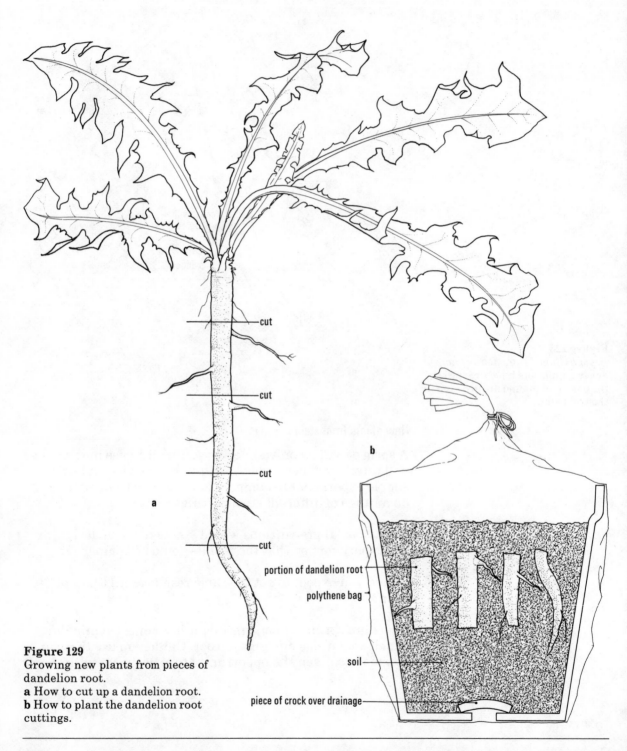

Figure 129
Growing new plants from pieces of
dandelion root.
a How to cut up a dandelion root.
b How to plant the dandelion root
cuttings.

Introducing living things

2 Cut the taproot in sections 2–3 cm long (see *figure 129a*). Keep them upright.

3 Fill a plant pot with moist soil and plant the portions of root so that they are just below the surface of the soil as in *figure 129b*.

4 Cover the pot with a polythene bag to prevent evaporation of moisture from the soil. You will probably not need to water the pot.

5 After two weeks take the polythene bag off and examine the surface of the soil in the plant pot.

Q7 Can you see anything growing?

6 Dig up the root cuttings.

Q8 Do you notice any signs of growth?

Q9 Why do you think dandelions are such difficult weeds to get rid of in gardens?

Study *figure 130* carefully.

Q10 Describe how many ways of reproduction you can see taking place.

Figure 130
The houseplant *Chlorophytum capense*.

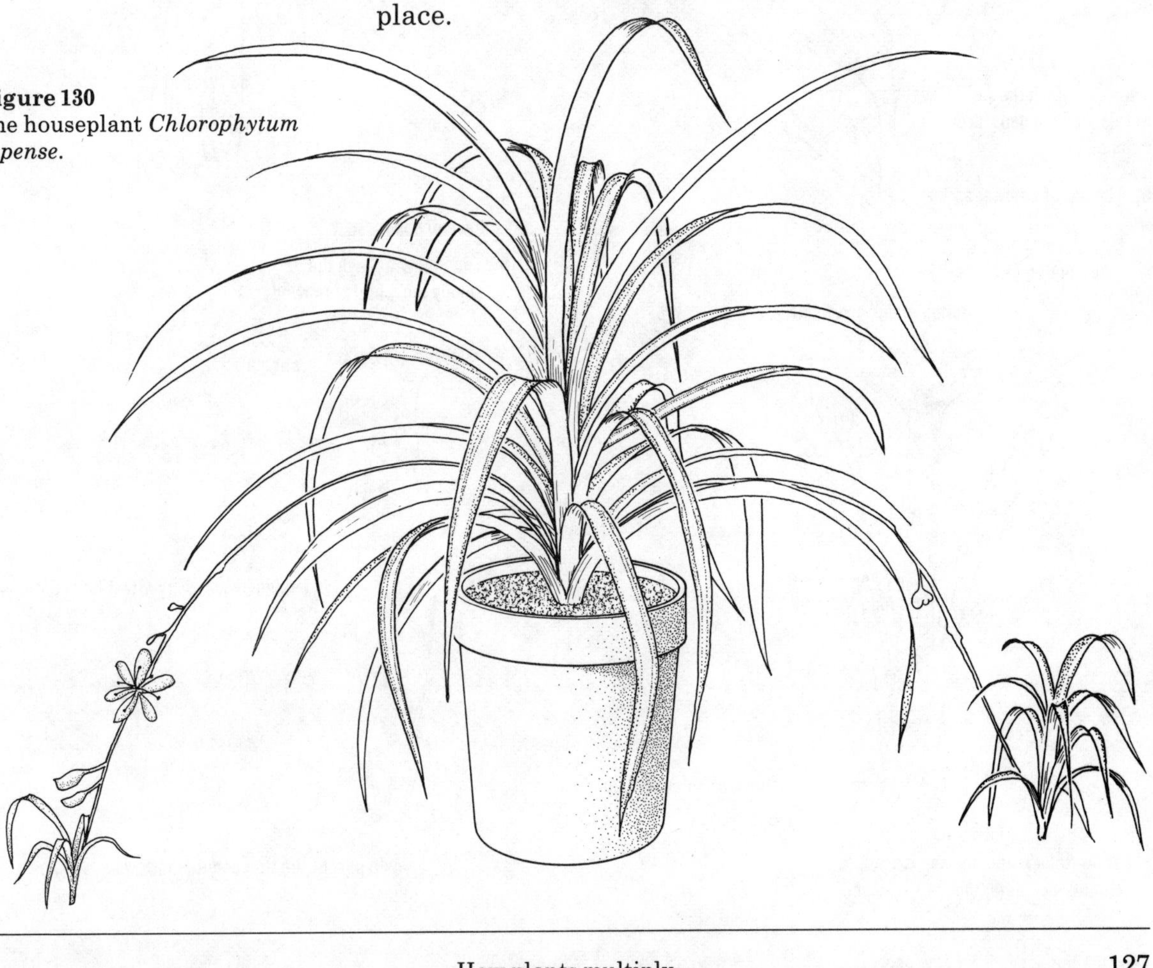

6.2 Looking at flowers

Most people have a clear idea in their mind about what a flower is. To begin with it is best to study at least one large flower to see the basic parts clearly. Later, flowers that are less obvious will be studied.

6.21 What are the different parts of a flower?

Choose, from the flowers provided, one in which the petals have just opened. Examine your flower and, using the drawings in *figures 131* to *133*, try to answer the questions. Also note the colour and texture of the different parts.

Figure 131
The different parts of a wallflower.

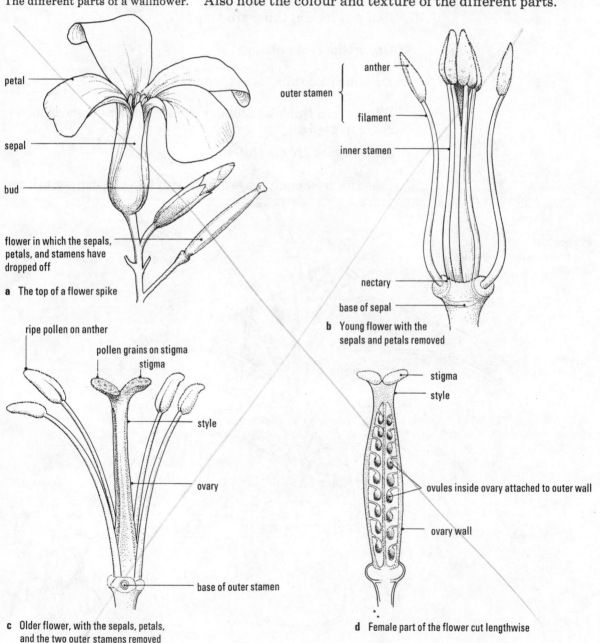

petal

sepal

bud

flower in which the sepals, petals, and stamens have dropped off

a The top of a flower spike

outer stamen

anther

filament

inner stamen

nectary

base of sepal

b Young flower with the sepals and petals removed

ripe pollen on anther

pollen grains on stigma

stigma

style

ovary

base of outer stamen

c Older flower, with the sepals, petals, and the two outer stamens removed

stigma

style

ovules inside ovary attached to outer wall

ovary wall

d Female part of the flower cut lengthwise

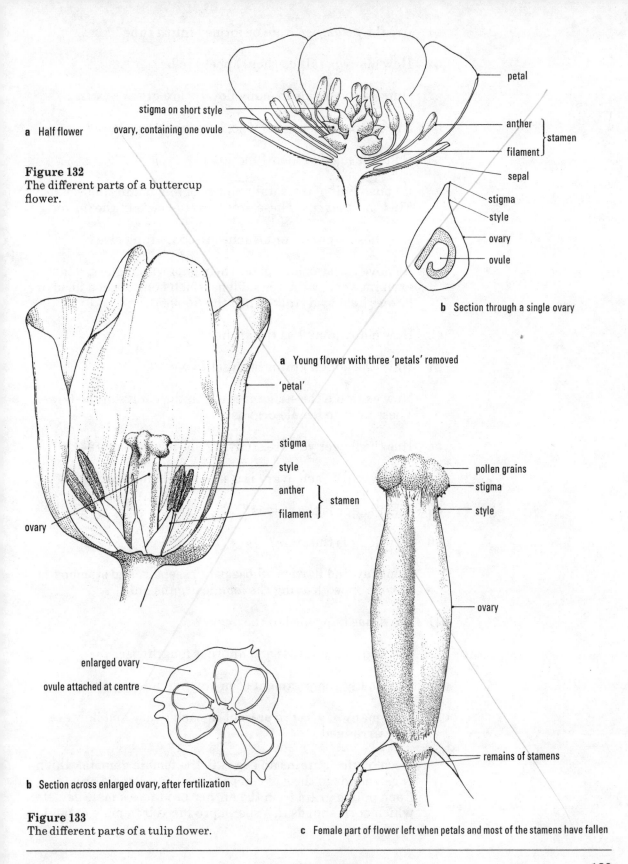

petal

stigma on short style

ovary, containing one ovule

anther

stamen

filament

sepal

a Half flower

Figure 132
The different parts of a buttercup
flower.

stigma

style

ovary

ovule

b Section through a single ovary

a Young flower with three 'petals' removed

'petal'

stigma

style

anther

stamen

filament

ovary

pollen grains

stigma

style

ovary

enlarged ovary

ovule attached at centre

remains of stamens

b Section across enlarged ovary, after fertilization

Figure 133
The different parts of a tulip flower.

c Female part of flower left when petals and most of the stamens have fallen

How plants multiply

LEAVE MIN 5 INCHES.

Underline
question

Italics =
CAPITALS

Q1   Are the *petals* separate or formed into a tube?

Q2   How many petals do there appear to be?

Outside the petals in many flowers are green *sepals*.

Q3   Are sepals present, and if so, how many are there?

Q4   Can you suggest their function?

Remove half of the total number of any petals and sepals.
Find the *stamens*. These are the male parts of the flower.

Q5   Are these separate or attached to something else?

Remove the stamens, place them on a watch glass, and
examine them with a hand lens. Each consists of a head or
anther held by a stalk called the *filament*.

Q6   How many lobes has the anther?

Q7   Have the pollen grains been set free yet?

Now examine the structures left in the centre of the flower.
These are the female organs.

Q8   Has the flower you are studying one *ovary* or more?

At the top of each ovary is a *stigma* held up by a *style*.

Q9   What shape is the stigma?

Q10   How long is the style?

You may find flowers whose petals, sepals, and stamens
have dropped leaving the female organs only.

Q11   What has happened to the ovary?

Cut one ovary across and another lengthwise.

Q12   How many compartments can you find in each ovary?

Q13   How many *ovules* are present in each ovary and how are
they arranged?

Each ovule contains an egg cell, the female gamete, which
corresponds to the egg cell produced by a female animal.
Each pollen grain from the anther produces a male gamete,
which corresponds to a sperm produced by a male animal.

6.22 Insects and flowers

You must have noticed insects visiting flowers.

Q14 Can you suggest why so many flowers are visited by insects?

Some flowers have coloured stripes and long rows of hairs on the petals.

Q15 Can you suggest what functions these have?

With the help of the drawings in *figure 134*, try to find the nectaries in the flowers you are looking at.

Q16 Do you think the same kinds of insects visit all the flowers shown in *figure 134*?

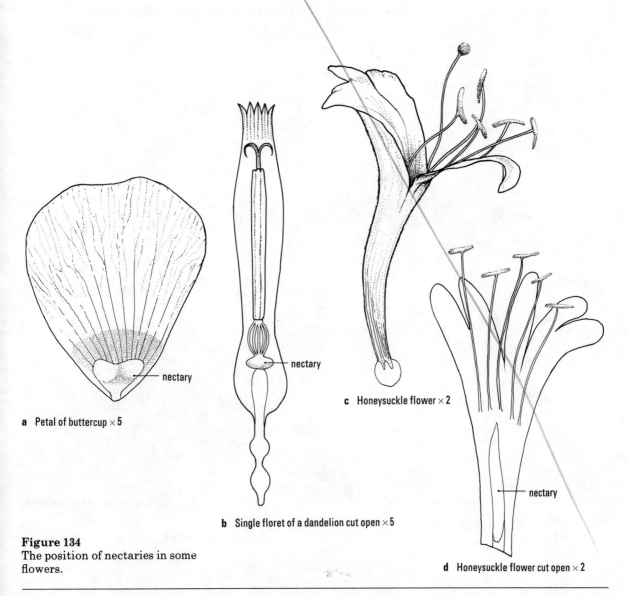

a Petal of buttercup × 5

b Single floret of a dandelion cut open × 5

c Honeysuckle flower × 2

d Honeysuckle flower cut open × 2

Figure 134
The position of nectaries in some flowers.

Some flowers also have strong scents as well as attractive colours. The flowers of the lime *(Tilia europaea)* smell strongly in the early morning, while others, such as the night-scented stock and honeysuckle, are scented more strongly in the evening. Some flowers, such as lupins, may be brightly coloured and yet have no scent. Such flowers have no nectaries.

Colour, scent, and guide lines all play a part in attracting insects.

Q17 Why does it seem to be important for flowers to attract insects and what, if anything, does the plant gain from attracting them?

Study *figure 135*, showing a bee collecting nectar.

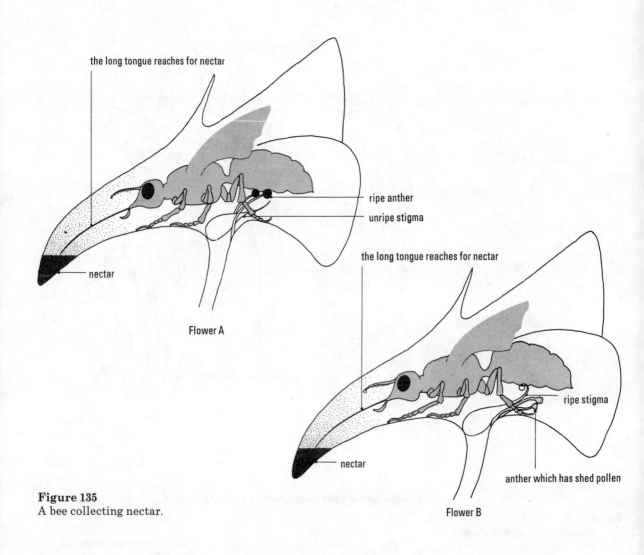

Figure 135
A bee collecting nectar.

Introducing living things

Q18   How can pollen be transferred from flower A to flower B?

The transfer of pollen from the anther to the stigma of a flower is called *pollination* and insects are important agents for doing this.

In some plants pollen is transferred from anther to stigma of the same flower; this is called *self-pollination*; one example is the garden pea. In other flowers *cross-pollination* is more usual.

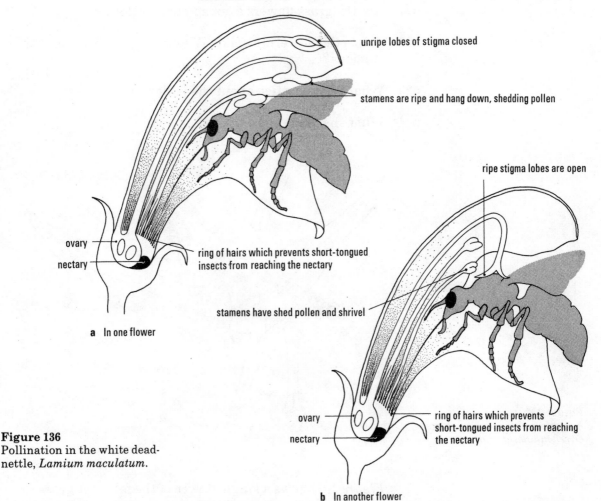

Figure 136
Pollination in the white dead-nettle, *Lamium maculatum*.

Q19   Which kind of pollination is illustrated in *figures 135* and *136*? Compare what is happening in the two flowers shown in each case.

Examine the pollen in the insect-pollinated plants provided, using a hand lens. Shake some of it onto a tile.

Q20   Does the pollen appear powdery or does it form small clumps?

6.23 Pollination by wind

There are many kinds of flowers which do not rely on insects to pollinate them but which use the wind to carry pollen.

Most grasses are pollinated by wind. By examining a grass flower you should be able to find out something about the special characteristics of wind-pollinated flowers.

Look at the flowering spike of a grass and, using a hand lens when necessary, try to answer the following questions:

Q21 Are the grass flowers a conspicuous colour?

Q22 Shake the flowering spike. What do you notice about the stamens?

Q23 What do you notice about the pollen?

Q24 What shape are the stigmas?

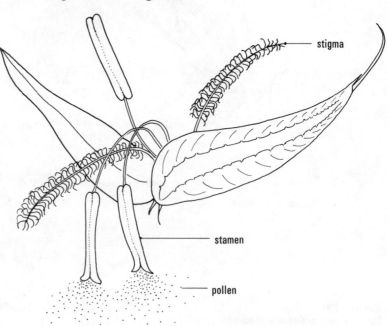

Figure 137
A single flower of false oat grass
(Arrhenatherum elatius).

Figure 137 shows a single flower of the false oat grass *(Arrhenatherum elatius)* with the stamens and stigmas.

If you have ever carried a twig of hazel catkins (*figure 138*) into the house in springtime, you will remember the thick dust of yellow pollen that settles on the furniture. On a windy day clouds of pollen can be seen as hazel catkins are blown about. The pollen grains are small and are produced in enormous numbers. Most wind-pollinated plants have pollen of this kind, which is able to float in the wind to reach the stigmas.

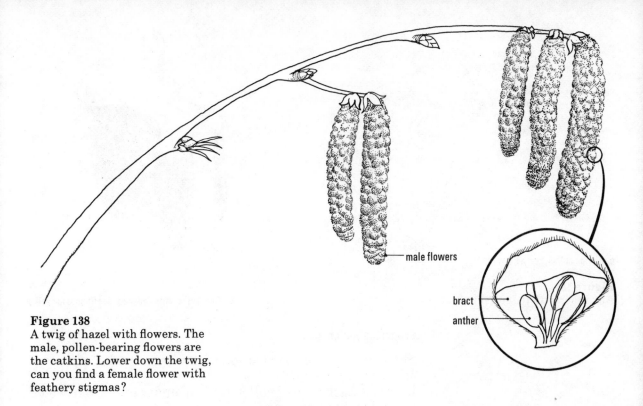

male flowers

bract

anther

Figure 138
A twig of hazel with flowers. The male, pollen-bearing flowers are the catkins. Lower down the twig, can you find a female flower with feathery stigmas?

Make a large copy of *table 11* in your notebook and complete it.

	Insect-pollinated flower	Wind-pollinated flower
Are the petals brightly coloured?		
Is nectar produced?		
Is there scent?		
Are the stamen stalks stout or slender?		
Is the pollen powdery or forming into clumps?		
What shape is the stigma?		

Table 11

6.3 Looking at pollen more closely

Figure 139 shows an anther with the top cut off before and after it is ripe. You may be given the opportunity to examine similar ones if these are available.

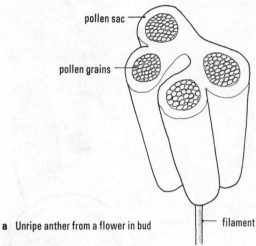

pollen sac

pollen grains

filament

a Unripe anther from a flower in bud

Figure 139
An anther with the top cut off.

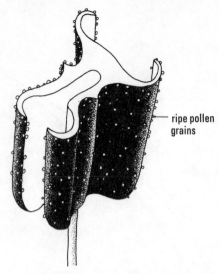

ripe pollen grains

b Ripe anther with pollen sacs split and ripe pollen

Examining pollen grains

1 Shake pollen from a selection of flowers provided onto a series of microscope slides.
2 Examine each type of pollen in turn, under low power, and if available, high power as well.

Q1 What colour is each kind of pollen?

Q2 What else do you notice about the appearance of each kind of pollen?

Study the three kinds of pollen shown in *figure 140*, parts *a*, *b*, and *c*.

Figure 140
Different kinds of pollen grains. These are all magnified about 500 times.

a The grains are smooth and sticky so that they easily adhere to the hairs on the insect's body

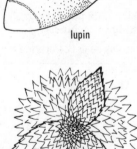

lupin

tulip

b The grains are covered with spikes which help them to cling to the hairs on an insect's body

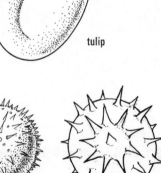

dandelion

hollyhock

sunflower

Introducing living things

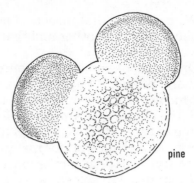

c Air-filled wings on either side of the grain help the pollen to float in the air

pine

Figure 140 *(continued)*

Q3 With which group, *a*, *b*, or *c* in *figure 140*, would you place each of the types of pollen grain you have examined?

Figure 141 *(right)*
Photomicrographs of different kinds of pollen, taken under high power.
a Pollen grain of the black pine *(Pinus nigra)* × 4575. The two air sacs make it light and easily blown about by the wind.
b Pollen of the common mallow *(Malva sylvestris)* × 170. The grains are covered with spines which help them to cling to the hairs on the body of an insect visiting the flower.
Photographs, R. H. Noailles.

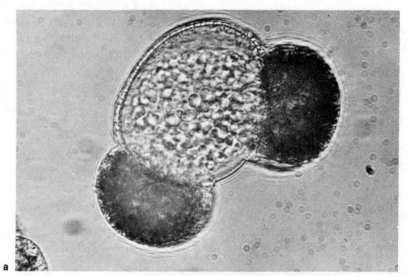

a

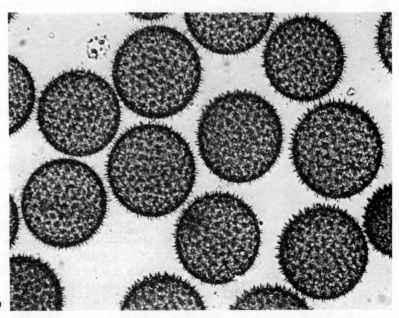

b

6.31 Growing pollen grains

We have seen that insects of many kinds carry pollen from one flower to another and that wind can also do the same.

Once the pollen has fallen onto the stigma it tends to stay there. But what happens next? How do the pollen grains, which contain the male gametes, reach the ovary, and in it, the ovules containing the female gametes? In some plants the style is very long. If you have examined a honeysuckle flower, for instance, you will have seen that the style can be 20 mm or more in length. Thus the pollen on the stigma will somehow have to travel this distance to reach the ovules at the other end. In order to do this a tube grows out of the pollen grain. This is called the pollen tube and we say that the pollen has *germinated*. *Figure 142* is a diagram of a pollen grain just starting to germinate and *figure 143a* is a diagram of germinating pollen grains on projections on the surface of the stigma. Now study the photomicrograph in *figure 143b* and compare it with *figure 143a*.

Figure 142
A pollen grain starting to germinate. The pollen tube grows down the style and eventually into the ovule of the flower. The male gamete travels down the pollen tube.

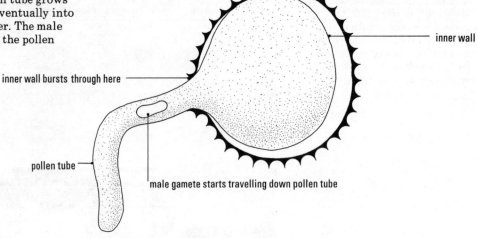

Q4 What are each of the structures labelled A, B, C, D in *figure 143b*?

What are the conditions needed for different kinds of pollen to germinate?

We know very little about what causes pollen to germinate, but the stigmas contain sugar, which is essential for the germination of pollen in many plants. Some kinds of pollen grow on a strong sugar solution, some on a weak sugar solution, and others require no sugar at all. There is now a chance for you to find out what strength of sugar solution different kinds of pollen require for germination. The class will probably be divided into groups for this investigation.

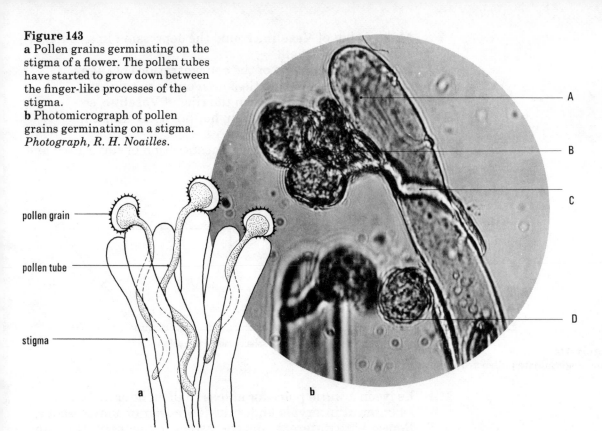

Figure 143
a Pollen grains germinating on the stigma of a flower. The pollen tubes have started to grow down between the finger-like processes of the stigma.
b Photomicrograph of pollen grains germinating on a stigma.
Photograph, R. H. Noailles.

pollen grain

pollen tube

stigma

a

b

A

B

C

D

The groups should try germinating pollen in the following solutions:

Distilled water
5 per cent sugar solution
10 per cent sugar solution
15 per cent sugar solution

Make a large copy of *table 12* in your notebook, ready for your results.

Pollen	Results √ if any pollen grains germinate × if no pollen grains germinate			
	Distilled water	5 per cent sugar solution	10 per cent sugar solution	15 per cent sugar solution
Bluebell				

Table 12

1 Make a ring of Vaseline round the depression in a cavity slide, forming a well.
2 Fill the well with one of the concentrations of sugar solution and shake a little pollen onto the solution.
3 Place a coverslip lightly on the ring of Vaseline, so enclosing the sugar solution, but not squashing the Vaseline.
4 Label the slide to show the concentration of sugar used and the type of pollen.

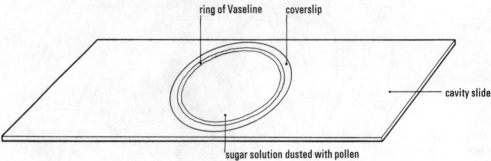

Figure 144
How to germinate pollen grains.

5 Leave in a warm place for at least half an hour and then examine at intervals under the low power of a microscope.
6 Repeat with different concentrations of sugar solutions and with pollen from other kinds of flower.
7 Record your results as shown in *table 12*.

By comparing your results with those obtained by the other groups, you can tell whether pollen from each of the flowers germinated best in distilled water or in one of the sugar solutions of different strengths.

6.4 How are seeds formed?

In *figure 143* we can see pollen tubes growing down the style towards the ovule of the flower. The pollen tube grows into the ovule, carrying with it the male gamete. This male gamete fuses with the female gamete inside the ovule to bring about fertilization. This process is equivalent to the fertilization of an egg by a sperm in animals. The fused male and female gametes develop into a small *embryo*.

This embryo gradually develops and grows inside a protective cover formed from the ovule. Together these form a *seed*. The ovary also enlarges and becomes a *fruit*, containing one or more seeds.

Study the flowers provided. Copy *table 13* into your notebook and complete it.

	Before fertilization	**After fertilization**
Sepals Petals Stamens Stigma Style Ovary Ovule		

Table 13

Looking at developing fruits

Open the developing fruits provided.

Q1 How many ovules have enlarged to become seeds?

Find an ovule which has not enlarged.

Q2 Why has this ovule remained small?

Fruit and seed dispersal

Study a selection of ripe fruits and seeds. Sort these into groups, each containing examples scattered or *dispersed* in the same way.

Some methods of fruit and seed dispersal are very obvious (see *figure 145*) and very efficient but other plants do not have any apparent features for carrying dispersal out.

Q3 Can you suggest how the unspecialized fruits and seeds of these plants may be carried away?

Q4 Why is fruit and seed dispersal essential?

6.41 Sowing ripe seeds

The seeds you buy in packets are dry and ripe. They have been collected in a previous year from their respective plants. Try sowing some seeds from packets, together with some which you have collected yourself. The teacher will discuss with you suitable seeds to use.

Each group could sow one kind of seed.

1 Place pieces of broken crock over the hole in the bottom of a plant pot.

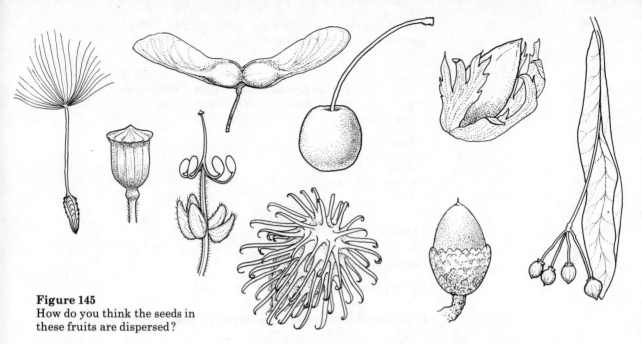

Figure 145
How do you think the seeds in
these fruits are dispersed?

2 Fill the pot with damp garden soil or seed compost to
within 1 cm of the top, pressing it down firmly as you do so.
3 Sow each large seed in a hole about 10 cm deep, made with a
pencil, and cover with soil but sow fine seed on the surface
and cover with a sprinkling of soil or sand.
4 Label the pot carefully with the name of the seed, your own
names, and the date.
5 Enclose the pot in a polythene bag and keep in a warm
place, along with the other pots prepared by the class.
6 Examine the pots regularly and keep the contents moist but
not wet.
7 Remove the covers when the seedlings are growing well
above the soil.

Keep the seedlings growing, either indoors or planted out
in the garden, if possible until they have produced flowers
and seeds once more.

Q5 Do all seeds germinate at the same speed?

6.42 Growth of seedlings

The seedlings obtained from the previous experiment or
similar ones can be used for this work.

Record the length of the seedlings at regular intervals and
plot the results on a graph of suitable scale.

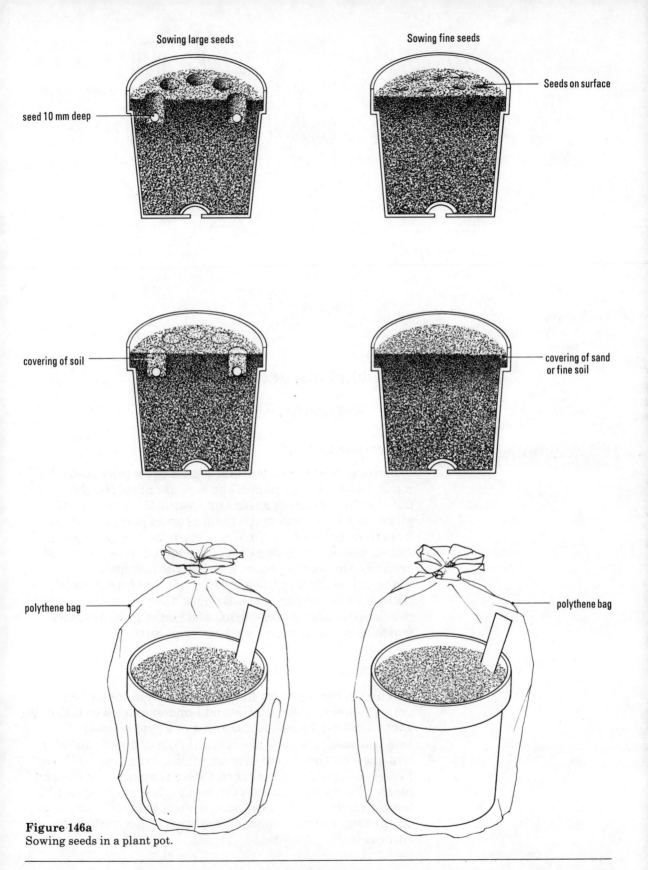

Sowing large seeds

Sowing fine seeds

Seeds on surface

seed 10 mm deep

covering of soil

covering of sand
or fine soil

polythene bag

polythene bag

Figure 146a
Sowing seeds in a plant pot.

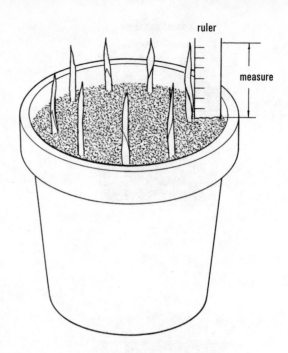

Figure 146b
Measuring the height of wheat seedlings.

Q6 Do seedlings all grow at the same rate?

Q7 Do seedlings grow at a steady rate?

Background reading

The plants in our lives

When man first found out that plants grow from seeds, he made a discovery as important as that of fire. He also found out that by collecting seeds and sowing them in another place, he could often make them grow in places far distant from their point of origin. Great changes in history have sometimes occurred because of the introduction of a new crop like that of the potato into Ireland and here we are going to look first at the activities of plant hunters and collectors in making these changes, then at where some of our familiar plants come from, and lastly at the ways in which man has bred new and better plants.

Plant collecting

The first recorded plant hunting expedition took place before Athens, Rome, Paris, and London existed as cities, in about 1500 B.C. Queen Hatshepsut of Egypt wanted frankincense, which is the scented gum of a tree, and this tree grew on the shores of East Africa, in the land of Punt. Five ships sailed and returned with quantities of seeds and plants. Thirty-one plants were established in the Queen's temple gardens at Karnak where, even today, the carvings on the temple walls, showing the trees in pots, mark the success of the expedition.

Figure 147
Queen Hatshepsut's plant
collectors.
By courtesy of Dr F. N. Hepper.

Since Queen Hatshepsut's time, men have continued to
collect plants for several reasons. The professional
gardeners travelled to collect ornamental plants to grow in
gardens, and the botanists wanted to increase their
knowledge of the many different kinds of plant species. But,
of course, the two groups overlapped in their interests and
the gardeners became botanists and the botanists
introduced plants.

The discovery of the Americas opened the widest field for
the discovery of new plants. By the seventeenth century the
potato, the tobacco plant, and hundreds of others grown for
their usefulness, their beauty, or their effectiveness as
medicines had been introduced. In the eighteenth century
botanists accompanied Captain Cook on his voyages to the
Pacific and they counted thousands of new kinds of plants.
You have heard of the famous mutiny on the *Bounty*. When
the mutiny took place the *Bounty* was on her way from the
Pacific to Jamaica carrying bread-fruit trees which, it was
hoped, would provide a new crop for the West Indies. Over
the past 150 years, plant collectors have explored the
Himalayas and distant parts of China and Tibet and many
endured appalling dangers in the quest. A story that
illustrates these dangers concerns the plant collector,
E. H. Wilson. He broke a knee in a rock avalanche while
searching for lilies in China in 1910. His men made a splint
out of his camera tripod as he lay helpless on a mountain
track. Just at that moment a train of more than fifty mules
appeared from the opposite direction. The track was too
narrow for the mules to be turned back and Wilson had to
trust to the sure-footedness of the mules as they stepped

over him. None of them trod on him and he survived. Afterwards he always walked with what he called his 'lily' limp.

Botanists are still sent all over the world from gardens such as the Royal Botanic Gardens at Kew, London, to study and collect plants. These gardens have also played their part in the spreading of plants into places far away from where they originally grew. A first-class example of this, which is also of economic importance, is the way the rubber tree was introduced into the Far East.

Many plants produce rubber, but the tree from which rubber is obtained on a commercial scale used to grow only in Brazil. In 1875, the botanist Henry Wickham was living at Manaos on the Amazon and he sent about 70 000 seeds of the tree to Kew. The seeds contain oil, which can evaporate, and therefore they have a limited life. To speed shipping and customs, Wickham labelled the cargo 'valuable botanical specimens'. At Kew, about 2000 healthy rubber plants were raised from the 70 000 seeds and they were then shipped to the Far East and became the parents of all the rubber trees in that part of the world, thousands of kilometres away from their country of origin. These provide most of the natural rubber needed nowadays.

The plants we eat and the plants we grow

Man's skill in moving plants successfully from one part of the world to another has had a great effect on our standards of living. Look in your kitchen at all the foods and drinks there that are made from plants. None of the grains we use is native to Great Britain and the wheat, barley, and oats we eat came originally from the Near and Middle East and the rice from the Far East. Maize and some of the beans we eat came from Mexico, and potatoes from the Andes. Many temperate and subtropical countries grow lemons, oranges, and limes, but all these came first from Asia. Tea is the dried leaf of a relation of the camellia. Most of it is grown in India, Sri Lanka (Ceylon), and East Africa, but it was introduced there from China and Assam. In the same way, coffee, grown in Africa, the West Indies, and Central and South America, is a bean from Arabia. Perhaps the sugar we sweeten it with comes from East Anglian sugar beet; but it is just as likely to come from one of the tropical countries that grow West Indian sugar cane, which in the first place came from India.

Go outside and look at your garden or a park and its surroundings. In the road outside, the horse-chestnut came, long ago, from the Himalayas and the laburnum from the Balkans; and that familiar city tree, the London plane, is a chance hybrid between a Persian and an American plane. The garden plants are even more cosmopolitan, with dahlias from Mexico, tulips from Turkey and Persia, daffodils (mostly) from central and south-western Europe, roses (in a tangled ancestry) from twenty or more countries including Persia and China, peonies probably from China, rhododendrons from Turkey, Tibet, the Himalayas, China, and the United States, and lilies from all the temperate countries of the Northern Hemisphere.

We owe much of the great variety of things we have to eat and drink and the wide range of flowers and trees we grow to the brave men and women who travelled in dangerous countries to find them. But we also owe much to another great series of discoveries. In this chapter you have learned about how plants reproduce. As men found out more and more about this subject, so they were able to breed new kinds of plants that suited their purposes better than any found wild.

Making new plants

During the seventeenth and eighteenth centuries, a new process, which we might call controlled plant breeding, entered the world of cultivated plants. For thousands of years the same process had occurred amongst plants but only by accident. Controlled plant breeding consists of man choosing plants which have desirable characteristics and then artificially transferring pollen from the stamens to the stigmas of the selected plants. The seeds which are produced are sown. If the plants which grow have the desired characteristics then these plants are propagated, often by one of the methods mentioned in this chapter, for example, cuttings.

The loganberry is thought to be a cross between the cultivated blackberry and the raspberry, which happened in California during the nineteenth century.

One of the most interesting as well as the most complicated examples of controlled plant breeding is that of the rose, which has gone on for many years. It has resulted in our modern hybrid tea and floribunda roses.

All our garden roses have been bred from a number of wild roses which have been interbred for hundreds of years. A new variety of a rose begins with the sexual method of reproduction you studied earlier in this chapter.

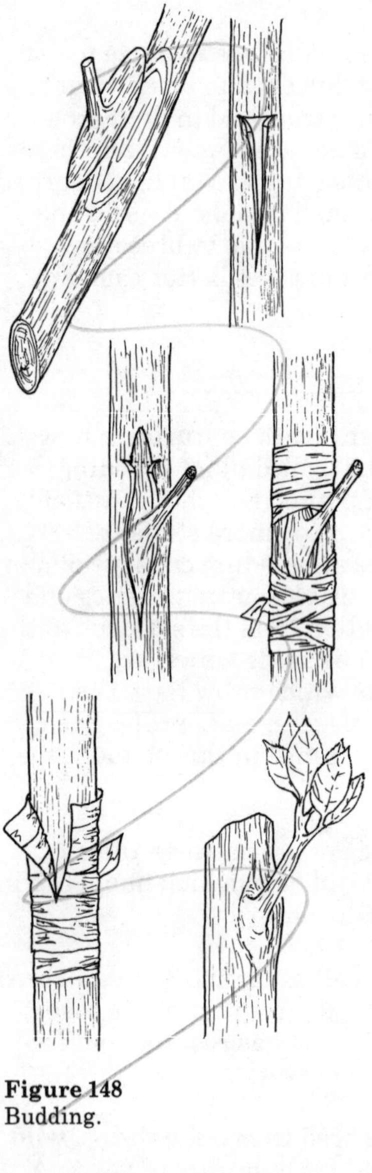

As roses have the stamens and the stigma in the same flower, self-pollination could occur by the pollen falling from the anthers onto the stigma. Cross-pollination could occur by insects carrying the pollen from one rose to another. A rose-breeder, however, does the pollinating by hand and his roses are usually grown under glass so that insects are kept out. The rose-breeder removes all the stamens from the two roses he wishes to breed from and he must do this before the anthers are ripe. He keeps the anthers until they ripen and shed their pollen and then he carefully transfers some of the pollen grains to the stigma of the rose he has chosen. He may use a small paintbrush, or his fingers to do this.

The breeder then watches to see if fertilization has occurred, by waiting, as the flower fades, to see if hips containing the seeds have formed.

These seeds are sown and as the seedlings come into flower, the rose-breeder is looking not only for a beautiful rose flower but for a plant free of such diseases as black spot and mildew (see *figure 169*).

To continue the breeding of a new rose, an asexual method of reproduction known as *budding* is used. This means taking a bud, with a small piece of stem attached, from the seedling and putting it into a T-shaped cut in the base of the stem of an *understock*, which is usually a wild rose plant. The bud is tied on and once the buds are growing on the understock the upper part of the understock plant is cut off. This means that the new rose is growing on the roots of a wild rose species.

When a new rose is put on the market, every plant has been directly grown from the buds of the first seedling of its kind – so sexual and asexual methods of reproduction play their part. Although wild roses will come from seed, cultivated ones rarely do so. They give, instead, a variety of seedlings. Therefore, 'budding' is necessary to ensure that all the roses of that variety breed true to form. One of the most popular roses, 'Peace', was bred by the French rose grower Meilland, and the millions of Peace roses throughout the world have all come from the buds of the original seedling which he grew.

Figure 148
Budding.

7

Man and microbes: the discovery of small organisms

In this chapter we are investigating small organisms. Some of them are so small that you need to look at them under a microscope to see them at all.

7.1 Leeuwenhoek and his 'little animals'

Here is a translation of some observations written by Antonie van Leeuwenhoek in 1676. You read about his microscope in Chapter 3.

'... 3. May the 26th, I took about ⅓ of an ounce of whole pepper and having pounded it small, I put it into a Thea-cup with 2½ ounces of Rainwater upon it, stirring it about, the better to mingle the pepper with it, and then suffering the pepper to fall to the bottom. After it had stood an hour or two, I took some of the water, before spoken of, wherein the whole pepper lay, and wherein were so many several sorts of little animals; and mingled it with this water, wherein the pounded pepper had lain an hour or two, and observed, that when there was much of the water of the pounded pepper, with that other, the said animals soon died, but when little they remained alive.

'June 2, in the morning, after I had made divers observations since the 26th of May, I could not discover any living thing, but saw some creatures, which tho they had the figures of little animals, yet could I perceive no life in them how attentively soever I beheld them.

'The same day at night, about 11 a clock, I discovered some few living creatures: But the 3d of June I observed many more which were very small but 2 or 3 times as broad as long. This water rose in bubbles, like fermenting beer.

'The 4th of June in the morning I saw great abundance of living creatures; and looking again in the afternoon of the same day, I found great plenty of them in one drop of water, which were no less than 8 or 10000, and they looked to my eye, through the Microscope, as common sand doth to the naked eye. On the 5th I perceived, besides the many very small creatures, some few (not above 8 or 10 in one drop) of an oval figure, whereof some appear'd to be 7 or 8 times bigger than the rest. . . .'

7.11 Can the same observations still be made today?

Now look at a sample of water from a pond or aquarium. The sort of microscope Leeuwenhoek used is shown in *figures 37* and *38*; the one you are using is rather different.

1 Use a dropping pipette to take a sample of the water you are going to examine.
2 Place a small drop in the centre of a microscope slide.
3 With the low power objective of the microscope in position, place the slide on the stage and look at the sample.
4 Record in writing, or by drawings, what you can see.

Q1 How many times have the objects been magnified?

5 If you see some small organisms in the sample of water, remove the slide from the stage of the microscope and place a coverslip on the drop of water.
6 With the low power objective still in position, return the slide to the stage of the microscope, look again at the sample, and then change to the high power objective.

Q2 How many times are the organisms magnified now?

7 Add any further observations to the record you have already made.

Q3 How do your observations compare with those of Leeuwenhoek?

Q4 What was remarkable about what Leeuwenhoek saw and why do we still remember him today?

7.2 What are microbes?

Because these organisms can only be seen under microscopes they are called *microbes* or *micro-organisms*. Beyond the fact that they are very small, what exactly do we mean by 'microbes'? Nowadays, the word 'microbes' is the name given to a group of organisms which includes such things as *1* the tiny fungi, *2* the one-celled animals called protozoans, *3* the very small bacteria, *4* the even smaller viruses.

Figure 149
Photomicrographs of microbes.
a Cells of yeast *(Saccharomyces cerevisiae)* (× 2000). Some are in the process of division.
Photograph, Unilever Educational Publications.
b The protozoan *Paramecium aurelia (* × 250).
Photograph, M. I. Walker.
c *Bacillus megaterium.* This is one of the largest bacteria. Each cell is 2.8 μ long and 1.2 μ thick. It is to be found in soil (× 5000).
Photograph from Hawker, L. E. et al. (1960) An introduction to the biology of micro-organisms. Edward Arnold.
d The tobacco mosaic virus (× 49000).
Photograph, Dr A. Klug.

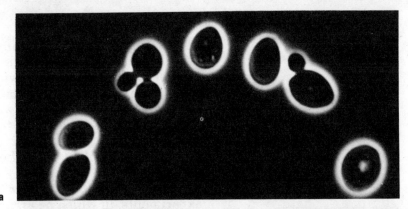

a

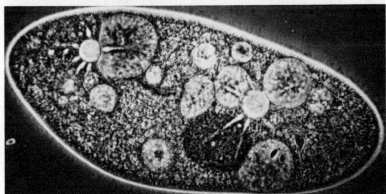

b

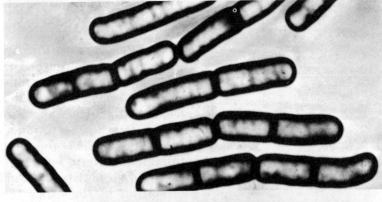

c

d

Leeuwenhoek carried out many different kinds of biological experiments, but in spite of his efforts he never got much further towards finding out what his 'little animals' really were or where they came from. Six years after his death an Italian, Lazaro Spallanzani, was born. (See *figure 161*.) Spallanzani was to become a famous scientist and his experiments served to continue the study of microbes more or less from the point where Leeuwenhoek left off.

In about 1770 Spallanzani carried out many now famous experiments, in which he arranged a number of flasks, each containing different liquids prepared from water with such things as maize, wheat, barley grains, beans, and peas.

Here is a translation of part of Spallanzani's own account of his experiments. Some of the more difficult words are explained below.

'I took nine vessels with seeds, hermetically sealed. I immersed them in boiling water for half a minute. I immersed another nine for a whole minute, nine more for a minute and a half, and nine for two minutes. Thus, I had thirty-six infusions. That I might know the proper time to examine them, I made similar infusions in open vessels, and, when these swarmed with animalcula, I opened those hermetically sealed.

'I examined the infusions, and was surprised to find some of them an absolute desert; others reduced to such a solitude, that but a few animalcula, like points, were seen, and their existence could be discovered only with the greatest difficulty. The action of heat for one minute, was as injurious to the production of the animalcula, as of two. The seeds producing the inconceivably small animalcula, were, beans, vetches, buckwheat, mallows, maize, and lentils. I could never discover the least animation in the other three infusions. I thence concluded, that the heat of boiling water for half a minute, was fatal to all animalcula of the largest kind; even to the middle-sized, and the smallest, of those which I shall term animalcula of the higher class, while the heat of two minutes did not affect those I shall place in the lower class.'

Spallanzani's animalcula of the 'higher class' were probably protozoans and the 'lower class' bacteria. 'Hermetically sealed' means that no air could get to the infusion from outside. An 'infusion' is made by grinding the seeds up with water.

Introducing living things

Spallanzani then carried out another experiment.

'I took sixteen large and equal glass vases: four I sealed hermetically: four were stopped with a wooden stopper, well fitted; four with cotton; and the four last I left open. In each of the four classes of vases, were hempseed, rice, lentils, and pease. The infusions were boiled a full hour, before being put into the vases. I began the experiments 11 May, and visited the vases 5 June. In each there were two kinds of animalcula, large and small; but in the four open ones, they were so numerous and confused that the infusions, if I may use the expression, rather seemed to teem with life.

'In those stoppered with cotton, they were about a third more rare; still fewer in those with wooden stoppers; and much more so in those hermetically sealed. . . . The number of animalcula developed, is proportioned to the communication with the external air. The air either conveys the germs to the infusions, or assists the expansion of those already there. . . .'

Spallanzani's experiments helped Louis Pasteur, who in 1861 carried out some further experiments, which have since become famous. There is more about Spallanzani's work in the Background reading.

Figure 150
Discovering where microbes come from.

Before reading about Pasteur's experiments, look at the results of some similar ones for yourselves. The tubes are set up as shown in *figure 150*.

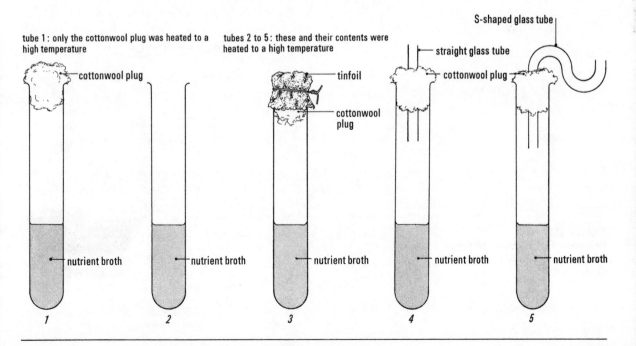

tube 1: only the cottonwool plug was heated to a high temperature

tubes 2 to 5: these and their contents were heated to a high temperature

cottonwool plug

tinfoil

cottonwool plug

S-shaped glass tube

straight glass tube

cottonwool plug

nutrient broth — nutrient broth — nutrient broth — nutrient broth — nutrient broth

1   2   3   4   5

Tubes 2–5 were heated to a high temperature in a pressure cooker or autoclave for 15 minutes. You will be told how long they have since been left and where they have been kept. It may be helpful to look at them again after a further period of time. Make a table of the results, recording the appearance of the broth.

	Day	Day	Day	Day	Day
Test-tube 1					
Test-tube 2					
Test-tube 3					
Test-tube 4					
Test-tube 5					

Table 14
Table which may be used for recording the results of the experiment described in *figure 150*. Insert the number of the day at the top of the vertical columns.

Q1 What conclusions are you able to draw from the results about
a the effect of a high temperature
b the effect of air coming into contact with the broth
c why the broth goes cloudy
d where the cause of the cloudiness comes from?

The S-shaped glass tube in test-tube 5, *figure 150*, is copied from Pasteur's flask. His actual flask is shown in *figure 151*.

Figure 151
Pasteur's famous culture flask.
Photograph, The Mansell Collection.

There is more about Pasteur and his experiments in the Background reading.

At the time when Leeuwenhoek, Spallanzani, and Pasteur lived, most people believed that living things could suddenly be formed from non-living things. This idea is known as the 'theory of spontaneous generation'. Leeuwenhoek made no suggestion as to the source of his 'little animals' but Spallanzani certainly began to question whether spontaneous generation did, in fact, take place.

In Spallanzani's time, experiments of the kind he did were quite new in science. People had been content to record and draw conclusions only from what they could observe under natural conditions. The idea of setting up a hypothesis and then designing an experiment to test it was only just beginning to dawn.

Q2 What do you think of Spallanzani's experiments, judged as scientific investigations?

Q3 If you were doing the same experiment today what improvements would you suggest, if any?

Q4 Was he right in drawing the conclusions he did draw, or would he have been justified in going further? (Note exactly what he did conclude.)

Q5 Spallanzani did not disprove the theory of spontaneous generation. After reading more about Pasteur's experiments, why do you think Pasteur succeeded in disproving the theory when Spallanzani did not?

7.4 What do microbes do?

You can find out quite a lot about microbes for yourselves: how they spread, and what effects they have on animals and plants and even upon ourselves. Some of these effects are bad, for many bacteria, viruses, and fungi can cause disease of one kind or another – whooping cough, measles, and ringworm, for example.

But by no means all microbes are harmful. Certain kinds which we shall be learning about are helpful in preventing diseases, in the preservation of food, and in the making of such things as cheese, wine, and bread.

We shall be considering microbes and their connection with disease later.

At the moment we will consider ways in which microbes are of use to man.

Bread-making

Most of you will already know that a baker uses yeast in making the dough for bread. Baker's yeast is a microscopic fungus and you can buy living yeast from a baker. In large quantities it feels rather like a lump of soft putty. If you wish to look at yeast under the microscope
1 make a suspension of a little of the 'solid yeast' in water
2 place a drop on a microscope slide
3 place a coverslip on the slide over the drop of liquid
4 look at the yeast first with low power magnification and then with high power.

Alternatively, you may be provided with a sample of yeast which has been stained to make it clearer.

Q1 How many times are the yeast cells magnified in each case?

How does yeast act in dough?
You will probably work in groups for this investigation and will be able to collect class results.
1 You will be provided with 3 different doughs, A, B, and C.
2 If you wish to study the effect of different temperatures on the dough you will need one set of tubes for each temperature, e.g. in the refrigerator, at room temperature, and at 37 °C.

Decide what temperatures you are going to apply *before* starting, and do all your tests at the same time.

3 Pour each dough, to a depth of about 3 cm, into 3 different boiling tubes or measuring cylinders labelled A, B, and C to correspond with the dough. Record the exact amount of dough.
4 Place 1 set of the tubes at each temperature selected.
5 Measure and record the height of the dough every two minutes or at shorter or longer intervals, if you think it more suitable, and make a table of the results.
6 When you have enough readings, draw a graph of the results.

You will be told what each dough contains.

Q2 What do you think causes dough to rise?

Q3 Exactly what part do you think yeast plays in causing dough to rise?

Brewing and wine-making

Man has also made use of the fact that other yeasts are able to ferment sugar solutions and make alcohol, which is a way of preserving a beverage. For instance, it is used in the brewing of beer and in the making of various wines from fruit juices and from parts of plants such as dandelion flowers. Most kinds of fruit, but particularly grapes, have yeast cells growing on their skins. So, when grapes are crushed to make wine, the yeast from their skins acts on the sugar they contain to make alcohol. The ancient Egyptians made a kind of beer 6000 years ago, and before the Romans brought barley to Britain, the ancient Britons made a beer using wheat.

To make beer, the brewer soaks barley grains in water in a warm place so that they start to grow. As they sprout, sugars are produced and they are then called the 'malt' grains. The brewer heats the malt grains and so prevents them from growing any further. He adds more water to the grains so that the sugar soaks out of them. This extract is called the 'wort'. The wort is boiled and the hops are added. Hops provide the 'bitter' flavour, although originally they were added as a preservative. When the wort has cooled the brewer adds the yeasts. He allows the whole mixture to ferment and this produces alcohol.

Cheese-making

Figure 152 shows two of the ways in which microbes are used to produce different varieties of cheese.

Figure 152
(Left) Emmental cheese. The holes are made by bacteria producing carbon dioxide. *Photograph, Swiss Cheese Union, Inc.*
(Right) A blue Stilton cheese. A mould is used to cause the greenish-blue markings. *Photograph, English Country Cheese Council.*

Figure 153
a A compost heap just being made.
Photograph, R. J. Corbin.

Yoghourt

Yoghourt has recently become a popular food in this
country although it has been eaten in many parts of
Eastern Europe and the Middle East for a long time.
Bacteria are the micro-organisms which ferment the milk
to produce yoghourt. *Lactobacillus casei* causes the
fermentation and another bacterium *Streptococcus
thermophilus* is added to give the creamy flavour.

Decay and disposal

Can you imagine what the countryside would be like if the
bodies of animals which died did not decay or if wood and
leaves did not rot away? This is perhaps the greatest value
micro-organisms have for man. The countryside would have

a very different appearance if microbes were not so efficient at changing dead vegetable and animal matter into harmless materials which become part of the atmosphere or soil. The ability of microbes to cause decay of organic matter can have disadvantages where food is concerned, but we must remember also how useful this ability is.

Man makes use of the effect of microbes on organic matter in sewage disposal. In this country, up to the nineteenth century, sewage accumulated in the streets of cities and many rivers were open sewers because rain washed refuse into them. The Public Health Act of 1876 and a Royal Commission in 1898 changed this and now, in the twentieth century, sewage disposal is traken for granted. Bacteria act on the sewage sludge and oxidize much of it away as the gas, carbon dioxide, which simply goes into the air. Another gas, methane or marsh gas, is also produced during the process and this gas may be used by the sewage works to drive their machinery.

Figure 153 (continued)
b An old compost heap.
Photograph, R. J. Corbin.

7.5 Where are microbes found?

Before you can study bacteria or other microbes and discover where they grow naturally, you need to know how to collect and grow them.

The first methods that were used to grow them are very like the methods used today. About the time that Pasteur had become famous for his experiments, a German called Robert Koch (1843 to 1910), whose photograph appears in *figure 154*, was busy experimenting in a small laboratory in his garden. He wanted to know how bacteria lived, and how they could be grown. Working completely on his own, Koch found ways of growing bacteria. Since he was interested in those found in diseased animals, he looked for methods of cultivating bacteria in such a way that he could watch their growth under laboratory conditions. So he infected a drop of liquid from the eye of an ox and placed the drop on top of a microscope slide in which there was a shallow cavity. He then turned the slide over so that the drop hung suspended (see *figure 155*). Under his microscope he watched bacteria grow, forming more and more bacteria. He even made the first photographs of them, through his microscope.

Figure 154
Robert Koch in his laboratory. He is looking at one of his cultures under the microscope.
Photograph, The Mansell Collection.

Introducing living things

Figure 155
Robert Koch's 'hanging drop'
slide.

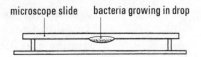

Later, Koch improved his methods of culturing bacteria by growing them on plates of jelly. This made it possible for him to study their shapes more easily with a microscope and also to select different kinds of bacteria from the jelly plates and to grow each separately. These are called pure cultures, and Koch's methods are still those used by bacteriologists today.

7.51 Microbiological techniques

You are going to make cultures of bacteria, using nutrient agar as a growing medium. This is really a kind of jelly containing beef extract and other substances upon which bacteria feed. The agar, which is extracted from seaweed, forms a jelly at room temperature.

It is important to observe strict rules of cleanliness while you are preparing the experiments. If you are going to grow bacteria in Petri dishes, or 'plates' as they are often called, you must make sure that there are no bacteria on the plates before you begin, otherwise you will not know exactly where the bacteria have come from.

Here are some rules which you should follow.
1 Always wash your hands immediately before you touch the Petri dishes.
2 Once the lid of the Petri dish containing the agar is lifted off, avoid draughts and do not breathe or cough nearby. Replace the lid as quickly as possible.

Special precautions to be observed when carrying out experiments with micro-organisms
1 As soon as the plates have been inoculated, each lid must be attached to its base with adhesive tape, which must never be removed.
2 If a culture is spilled accidentally, report it to the teacher *at once* so that it may be cleaned up immediately and with complete safety.
3 After you have finished with a culture, dispose of it by placing the unopened Petri dish in a container of disinfectant or as directed.
4 Avoid all hand to mouth operations, such as eating food, until you have washed your hands.
5 Wash your hands, preferably before leaving the laboratory or immediately afterwards.

Recording your experiments
In all of these experiments, you will be using several Petri dishes and inoculating them in different ways. It is important, therefore, to label the *bottom of each plate* as it is inoculated. You will then know which is which when it comes to recording results. Keep careful records so that you can look at them before answering questions and drawing your conclusions.

7.52 Methods of collecting samples of micro-organisms

According to the type of place in which you are looking for micro-organisms, you may need to use different methods for collecting them.

These are some of the methods you might use:
1 Expose a sterile Petri dish to air in a chosen place, for a given time, by lifting the lid and then replacing it.
2 Lift the lid of the Petri dish, place the object you are investigating (such as your finger) directly on the sterile plate, wipe it across the agar, and replace the lid.

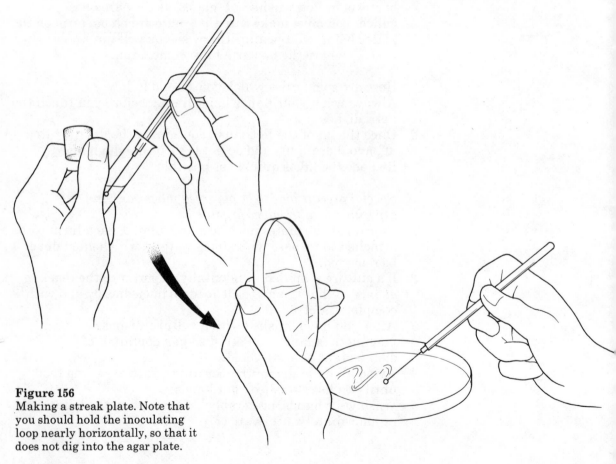

Figure 156
Making a streak plate. Note that you should hold the inoculating loop nearly horizontally, so that it does not dig into the agar plate.

3 Sterilize a wire inoculating loop by heating it in a Bunsen flame. Allow it to cool in the air without touching the bench. Use the sterile loop to collect your sample, from a liquid such as water or milk. Lift the lid of the Petri dish and smear the loop across the sterile agar and replace the lid. This is called a streak plate.

4 Mix the sample you wish to test, for instance, soil, with sterile distilled water and pour this over the plate, flooding it. Leave for 30 seconds, pour it off, and replace the lid. You can pour other liquids directly onto the plate in this way.

Attach the lids of the Petri dishes by sealing them with adhesive tape as soon as they have been inoculated. (See *figure 157*.)

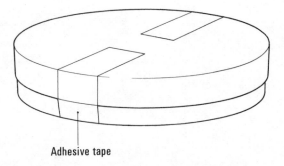

Figure 157

Adhesive tape

There are many places where you can look for bacteria. Some suggestions are water, soil, air, food, ourselves, anything you handle.

1 To enable you to discover just where bacteria are found you will be provided with sterile Petri dishes containing sterile agar.

2 You must remember the rules and special precautions and get well organized before you start.

3 You must also decide where you will take your samples and label the Petri dishes accordingly.

4 You must then decide the most appropriate method for collecting the samples.

5 When you have taken the samples and inoculated the plates, place them in an incubator at 37 °C.

6 Examine the plates in your next lesson.

7 Record the results. You will need to decide the best way to do this.

Q1 Is it possible to estimate the number of bacteria that were present in the different samples?

Q2 Is it valid to compare the number of bacteria in the different samples?

Q3 Write a summary of

a what you have discovered about where bacteria are found (use what you know of the results of others in the class as well as your own results);

b what you have learned about where bacteria come from.

7.6 Further facts about microbes

7.61 Bacteria in millions

Figure 158 is a graph of the rate at which bacteria increase, theoretically.

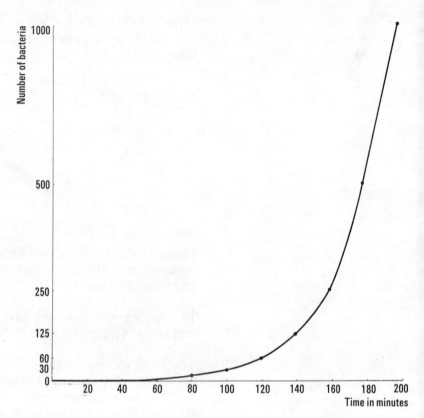

Figure 158
A graph based on theoretical figures, showing the rate at which bacteria increase.

Table 15 shows the actual figures for the growth of a population of bacteria in a culture, obtained by counting them over 36 hours.

Q1 According to the graph based on theoretical figures (*figure 158*), how long does it take for bacteria to double their numbers?

Q2 Compare the actual figures with the graph. How would a graph of the real figures differ from the graph of theoretical figures?

Hours	Number of bacteria
0	1 000
2	1 500
4	10 000
6	1 100 000
8	120 000 000
10	500 000 000
12	560 000 000
14	630 000 000
16	630 000 000
18	250 000 000
20	50 000 000
24	130 000
30	130 000
30	250
36	25

Table 15

Q3 Suggest what may be causing the change in the population after fourteen hours.

Q4 Suggest ways in which man can prevent bacteria from increasing.

7.62 Different kinds of bacteria

Figure 159 shows drawings of different kinds of bacteria.

Q5 Make a key of bacteria from the drawings in *figure 159*.

Q6 Three types of bacteria are shown in the photographs in *figure 160*. Name these types in your key.

Figure 159
Different kinds of bacteria.

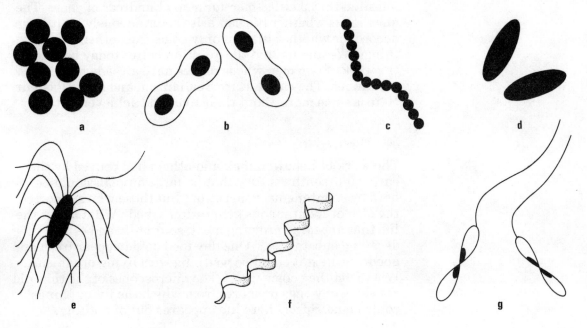

a

b

c

d

e

f

g

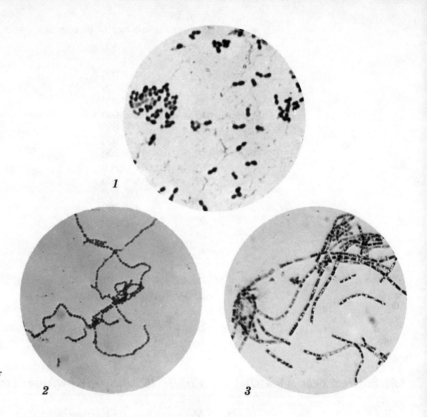

Figure 160
*1 Diplococci.
Photograph, Harris Biological
Supplies Ltd.
2 Streptococci.
3 Bacilli.
Photographs 2 and 3 by courtesy of
the Wellcome Trustees.*

Background reading

Life from nothing or life from life

The experiments you have done in the course of this chapter
will have told you a great deal in a short time about a
question that puzzled men for many hundreds of years. The
question is whether life can arise spontaneously (of its own
accord) or whether life can only come from other living
things. We understand the question better today because of
the work of two men whose names have already been
mentioned. They are Lazaro Spallanzani and Louis Pasteur.
Here is some more about their lives and achievements.

Spallanzani

The work of Leeuwenhoek and others had proved by the
early eighteenth century that no large animals could be
born by spontaneous generation. But those who supported
the idea of spontaneous generation asked, 'What about the
life that appears in rotting meat, sour milk, and other
decaying substances?' One day food could appear fresh and
good; the next it could be writhing with living organisms.
Where did they come from? The microscopes of the time did
not show any eggs or spores from which the living things
could come. Surely here life appeared out of nothing.

Introducing living things

Lazaro Spallanzani was born in Italy near Modena in 1729. He studied at the University of Bologna where a woman cousin of his was professor of physics. He became a priest and eventually was placed in charge of the museum at Pavia by the Empress Maria Theresa who ruled that part of Italy. He travelled widely in the Mediterranean, even visiting Turkey to collect natural history specimens for the museum. He was fascinated by volcanoes and made surveys of Vesuvius and Mount Etna.

Figure 161
Spallanzani and Mount Etna.
The Mary Evans Picture Library.

In common with many scientists of his time, he studied a great number of subjects. He wanted to find out about the digestion of food and experimented first on a vomiting crow, studying what it brought up after various lengths of time, and second on himself, swallowing bags and tubes of food, in order to find out what happened to the food. In another series of experiments he showed for the first time that artificial insemination was possible in a frog, a tortoise, and a bitch – a discovery that was much later to have a profound effect on the breeding of farm animals.

Spallanzani liked to try out experiments done by other people to see whether his findings fitted with theirs. In 1748 an English priest, John Needham, performed an experiment which he thought proved that organisms could be formed by spontaneous generation. Needham boiled some mutton broth and put it in a jar, taking especial pains to seal the jar, so that no one would be able to say that living organisms had been brought in by the air. When he opened the jar a few days later it was full of organisms. He tried the experiment again with other substances and always got the same result.

Spallanzani, hearing of Needham's work, tried the experiments himself twenty years later. It struck him that as the organisms were so small and could barely be seen under the microscope, then their eggs or spores must be so small they must be invisible. First he boiled seeds in water, sealed them, and opened them after a few days. He found as Needham did that they were full of life. Next he tried excluding air altogether. He took two batches of jars, sealed one batch with a blow pipe and boiled it, and left the other batch (the control group) open to the air. This time the first batch had very few organisms in it while the other was full of them. Now he tried boiling the flasks for half an hour, instead of for a few minutes. The result showed no organisms when the flasks were opened. From these experiments Spallanzani decided that where organisms had appeared in Needham's and his own earlier investigations they had come from invisible eggs or spores on the walls of the flasks or, where the flasks were open, from the air. Some of these organisms could withstand being boiled for a short time but none could live through the prolonged boiling that Spallanzani subjected them to.

Some were convinced by Spallanzani's arguments. Others said, no wonder nothing living appeared in the solutions boiled for a long time because this prolonged boiling killed off the delicate 'vital principle' in the air from which life came. It was left to Pasteur to settle that and many other arguments.

Spallanzani's work did have one practical result. The Emperor Napoleon wanted to make his armies as efficient as possible and to make sure they were never held up for lack of food. When he was still a general in 1795, he offered a prize for the invention of a method of preserving food. In 1810 a chef called François Appert won the prize with a method of heating food and then sealing it from the air. This was an application of Spallanzani's work, depending on the fact that spontaneous generation does not take place and therefore food does not putrefy. From Appert's discovery stem the canning industry and all the changes that industry has brought to our habits of eating.

Pasteur

The Emperor Napoleon enters this next story in another way. Pasteur's father had been a sergeant in Napoleon's army and the young Louis Pasteur grew up with a passionate love of his country and a pride in the glory Napoleon had brought to it. He too wished to do something great for his country and this desire urged him on to his great discoveries. Another deep force in Pasteur was his religion – he was a devout Catholic and this too had its effect on his outlook.

Figure 162
Louis Pasteur.
Photograph, The Mansell Collection.

Born in 1822, in the Jura, Pasteur did rather badly at school and wanted to be a painter. Some lectures by the chemist Dumas fired his interest in chemistry and made him work extremely hard. At the age of twenty-six he solved a problem that had puzzled some of the greatest scientists of the time. This was to do with a strange effect of light produced by tartaric acid (which is found in wine dregs) and in related substances. Pasteur produced an explanation which helped later chemists to find out far more about the structure of a great many substances. A famous and elderly scientist called Biot had been working on the same problem for many years. Pasteur had the frightening task of demonstrating his experiments to the old man who at the end said he was convinced by them and exclaimed, 'My dear child, all my life I've so loved this science I can hear my heart throb with joy!'

Pasteur got a job at Lille where he became interested in fermentation. Winemakers and brewers suffered great hardship from time to time because for no known reason their wine and beer went sour after ageing. A manufacturer of alcohol from sugar beet approached Pasteur in 1856 for his help over this very problem. Using his microscope on good samples of beer and wine and on soured samples, Pasteur saw that in the good samples the yeast was present as round cells but in the sour samples there were also some long cells present. From this he showed that there were two kinds of micro-organisms. The one with round cells was a yeast that produced alcohol; the other with long cells was a a bacterium and made lactic acid, the cause of the souring. He also showed that fermentation was the work of living organisms – this when the great scientists of the day had said it was caused by dead ones.

In the next few years he worked out a cure for the problem. He told the winemakers and the brewers that once the wine or beer was formed they should heat their products gently to kill off any yeast cells present, including the ones that produced lactic acid. The winemakers were horrified at the idea of heating wine but Pasteur showed by experiments with controls of unheated wine that the heated wine, once stoppered, did not sour, while several samples of the control group did go sour. His reputation grew and grew.

From his experiments on fermentation and yeast cells, he had become interested in finding out how micro-organisms arise. The question of spontaneous generation had been raised in public again. Reports appeared describing experiments that showed that life could arise out of dead matter. Supporters of spontaneous generation criticized the experiments of Spallanzani by saying that he had heated the air above his broths so much he had destroyed the vital principle in the air that could create life. The chief exponent of this idea was a doctor called Felix Pouchet, a highly religious man who collected pure air from mountain tops and devised ingenious methods of purifying air in order to use it in experiments to show that life could come from air alone. He supported spontaneous generation because he saw it as the means whereby God continues to create life.

Now Pasteur was equally religious. But he believed that God created life once and for all and provided it with its own means of continuation. He also had the results of his own experiments on fermentation in his armoury and a determined way of expressing his opinions. Pouchet tried to show that there were no micro-organisms in the air. Pasteur filtered air through guncotton and showed, by all the microbes left on the guncotton, that micro-organisms *were* in the air. Experiment and counter-experiment were tried but the crucial moment came when Pasteur and Pouchet were asked to put their points of view to the Academy of Sciences in Paris. Pasteur used the swan-necked flasks (see *figure 151*) to demonstrate that decay is caused by micro-organisms from the air. The broth was placed in these flasks, whose drooping necks trapped the bacteria from the air in their curves before they could get to the broth. The broth only went sour when the tops of the flasks were broken off. This is the same experiment, of course, as that described in section 7.3. Pasteur also cleverly managed to show that the air above the broth was not heated as in Spallanzani's method and that therefore no one could say that the vital principle in the air was harmed.

Scientists in France and England repeated Pasteur's experiments and most of them came to the same conclusions. Spontaneous generation was a wrong idea; life comes only from life. Out of all these experiments and arguments came a new idea, the germ theory of disease, something you will learn more of in the next chapter.

Microbes and health

8.1 Can we control the spread of microbes?

Most people will connect microbes with disease, although
many microbes are beneficial and useful to man and many
more are quite harmless. Disease-causing microbes are, in
fact, normally in the minority. You have seen how
widespread micro-organisms are, and how quickly they
multiply. It is obviously desirable to control the spread of
the microbes which cause food to spoil and those which
cause infection and disease.

Figure 163
A surgical operation in the 1870s.
*Radio Times Hulton Picture
Library.*

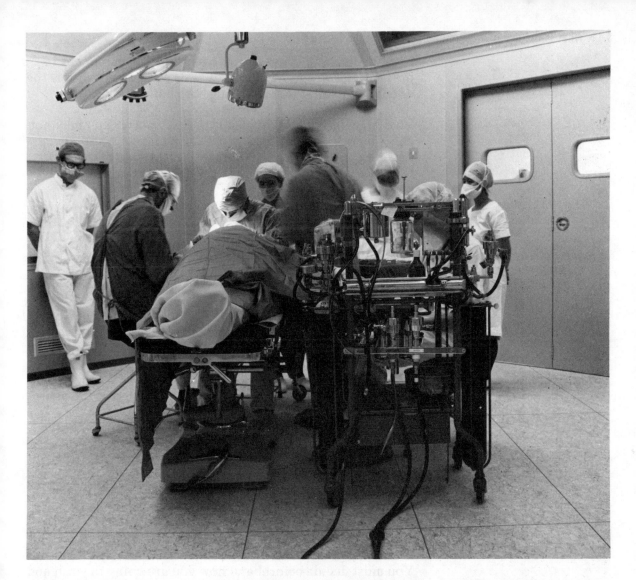

Figure 164
A modern operating theatre.
*Photograph by courtesy of
Moorfields Eye Hospital.*

In the mid-nineteenth century it was still quite common for people to die in hospital after an operation because of septic wounds. This is very unlikely to happen today.

Study the pictures of the two operating theatres in *figures 163* and *164* and answer the following questions.

Q1   In each operating theatre shown, can you see what provisions have been made to prevent the spread of infection during the operation?

Q2   What other precautions to prevent the spread of infection do you think may also have been taken in each theatre?

Q3   What sources of infection are present in *a* the operating theatre of the 1870s? *b* the modern operating theatre?

8.11 What is the use of washing (or scrubbing) the hands?

Most people will suggest that cleanliness is important in preventing the spread of microbes. Is this so and what is the effect of washing (or scrubbing) the hands?

This investigation is divided into three parts. You may not carry out all of them. You will be provided with the necessary sterile Petri dishes containing agar. Remember the rules which must be applied when carrying out this type of investigation.

Part I
For this part two people are needed to touch agar plates. One should *not* wash his hands beforehand.

1 Label three Petri dishes A, B, and C. (Remember to do this on the base.)
2 *The person with washed hands* lifts the lid of Petri dish A while the person *with unwashed hands* very quickly wipes the fingers of one hand lightly over the agar. Replace the lid quickly and seal it with adhesive tape.
   The person *with washed hands* lifts the lid of B and wipes the fingers of the other washed hand lightly over the agar. Replace the lid quickly and seal it.
4 Leave the Petri dish C unopened, but seal it with adhesive tape.
5 Place the Petri dishes in an incubator at 37 °C and examine them during your next lesson.
6 Record the results in the most appropriate way.

Part II
In this part of the investigation you will attempt to find out if washing your hands more than once makes any difference. You must decide beforehand how you are going to wash and dry your hands and do it the same way each time.

1 Take one Petri dish and before opening it, take a Chinagraph pencil and draw lines on the base dividing it into four sections.
   Label the sections as follows:
   '× 0' for unwashed hands
   '× 1' for hands washed once
   '× 2' for hands washed twice
   '× 3' for hands washed three times.
3 A person with washed hands lifts the lid of the Petri dish and one person places the finger of an unwashed hand on the section marked × 0. Replace the lid quickly.
4 The same person must wash his hands once and dry them in the manner decided.
5 Lift the lid of the Petri dish and place the finger of the washed hand in the × 1 section and replace the lid quickly.

6 Repeat this procedure after washing the hands twice and three times, placing the finger in the $\times 2$ and $\times 3$ sections respectively. Seal the Petri dish.

7 Place the Petri dish in an incubator at $37\,°C$ and examine it during your next lesson.

8 Record the results in the most appropriate way.

Part III
This is an opportunity for you to devise an experiment, using similar techniques to those in Parts I and II of this investigation.

Plan exactly how you would investigate:
whether drying your hands after washing them makes any difference and
whether the sort of towel you use makes any difference;
whether it makes any difference if you use a soap
b detergent c nothing;
whether it makes any difference if you scrub your finger nails or not.

These would be important points for food workers, nurses, doctors, etc.

Q4 What is the purpose of Petri dish C in Part I of the investigation and the section marked $\times 0$ in Part II?

Q5 Did you have a comparable Petri dish in the investigation you designed yourself? What was it?

Q6 You are always being told to wash your hands after using the lavatory. Can you now explain why?

8.12 Are disinfectants worth while?

If you cut your finger it will usually heal quite quickly, but sometimes it turns septic. 'Septic' really means going bad and until about a hundred years ago the reason why wounds went septic was not known.

You have already seen how one person's work often helped another to make an important discovery and here is yet another example. In 1865 an account of Pasteur's work showing that microbes in the air could turn broth bad reached an Edinburgh doctor, Joseph Lister. Lister argued that if microbes from the air could turn broth bad, could they not in a similar way turn wounds 'bad' after an operation? The old-fashioned method of treating a patient who had a leg cut off was to cover the wound with tar. This probably gave Lister the idea that pure carbolic acid, which occurs in tar, might kill the bacteria settling on the wound from the air. He tried soaking the dressings in carbolic acid and applied them to wounds immediately after an operation and he even had a spray of carbolic acid directed on the wound during the operation. (In *figure 163*, you can see Lister's spray being used.) This was successful and after two years of experimenting, Lister, in 1867, announced his discovery that carbolic acid applied to wounds would prevent them from going septic. Chemicals which kill germs are called *disinfectants*.

The term *antiseptics* is used for certain chemicals which prevent septic conditions developing in the body.

Carbolic acid is an unpleasant liquid to handle and Lister soon discovered that, while it certainly killed bacteria when used undiluted, it also damaged the tissues of the body. As a result, wounds treated with it refused to heal, or only did so very slowly. The problem facing Lister, then, was to find the strength of carbolic acid which kills bacteria but allows the growth of tissues over a wound.

One question you might ask in connection with the use of disinfectants is, 'Does the strength of the disinfectant used make any difference?' This question can be investigated experimentally.

Carbolic acid is distilled from coal tar and contains a substance called phenol. In the past, phenol has often been the most important ingredient of other disinfectants used in hospitals and laboratories, although other chemicals are now used instead. We shall therefore use another chemical substance which is a disinfectant but does not harm the tissues of the body.

Does the strength of the disinfectant used make any difference?
The following materials will be provided:
Petri dishes containing agar which has already been inoculated with bacteria called *Escherichia coli*, different dilutions of a disinfectant, full strength, $\frac{1}{10}$ strength, and $\frac{1}{100}$ strength, and distilled water.

1 Label four Petri dishes A, B, C, and D.
2 Pour full strength disinfectant over the Petri dish labelled A, leave it for 30 seconds, and pour it off. Replace the lid and seal with adhesive tape.
3 Repeat this procedure with the other Petri dishes but add $\frac{1}{10}$ strength disinfectant to B, $\frac{1}{100}$ strength disinfectant to C, and distilled water to D.

4 Place the dishes in an incubator at 37 °C and examine them in your next lesson or the following week.

5 Estimate the percentage of each Petri dish that is covered by colonies of bacteria and record it as shown in *table 16*.

Strength of disinfectant	Percentage of dish covered with a growth of microbes
A Full strength	
B $\frac{1}{10}$ strength	
C $\frac{1}{100}$ strength	
D none (distilled water)	

Table 16
Table for use in recording the effects of different strengths of disinfectant.

Q7 What are you able to conclude about the effectiveness of different strengths of disinfectant?

Q8 What other factors would you take into consideration when deciding
a   what strength of disinfectant to use?
b   which kind of disinfectant to use?

8.2 How can we prevent food from spoiling?

Perhaps it is not surprising that food which is good for us also contains the things that bacteria require for their growth. Obviously we want to stop food from going bad if we possibly can.

8.21 Pasteur discovers why wine can go sour

During the last century the wine growers in France found that the wine turned sour in their storage vats after a time, especially in warm weather. It had started to turn into vinegar. They asked Pasteur to investigate the trouble. He set up a makeshift laboratory in an old café in one of the wine growing districts of France. Taking samples of wine, he surprised the local wine growers by telling them which samples were going to turn bitter and which would remain good. The wine growers thought of Pasteur as a magician but there was really nothing miraculous about it. He simply examined the different samples of wine under a microscope and made his forecasts from the sorts of bacteria and the numbers that he found. (The Background reading for Chapter 7 gives some further details about this.)

After several attempts to solve the problem of how to stop the action of bacteria in wine, Pasteur finally hit on the idea of heating it to a temperature of 60 °C and then cooling it again. This killed off many of the harmful bacteria and made it possible to keep the wine almost indefinitely. Today this process is called *pasteurization* after Pasteur.

You also know that milk can go sour and you may have discovered through your investigations that bacteria are present in milk. So perhaps Pasteur's work on preventing wine from going sour can be applied to keeping milk 'fresh'.

Milk taken straight from the cow will be about the same temperature as the body of the cow – that is, about 41 °C. The first thing a dairyman does after milking is to cool the milk by passing it through a special cooling machine before it is put into churns.

Q1 Why is the milk cooled on the farm before it is put into churns?

Q2 How should the churns of milk be kept at the farm while waiting to be collected by the dairy?

8.22 How can we improve the keeping qualities of milk?

Milk bought from a dairy or any other shop is either pasteurized or sterilized. Commercially, milk is pasteurized either by heating it to 63 °C and keeping it at that temperature for 30 minutes, or by heating to 72°C for 15 seconds.

Milk is sterilized by heating it to 140 °C for 1 to 2 seconds.

It is possible to investigate whether these two different treatments do affect the keeping qualities of milk.
1 Label six sterile McCartney bottles A and label another six B.
2 Pour pasteurized milk straight from the bottle or container into the six labelled A.
3 Pour sterilized milk straight from the container into the six bottles labelled B and put the caps on immediately.
4 Place two bottles from each batch, that is, two labelled A and two labelled B, in a refrigerator, two from each batch in a warm place (you may use an incubator or stand the bottles above a radiator), and two from each batch out of doors.
5 Label each bottle with the place in which you have put it.
6 Leave the bottles until the next lesson and then take one of the bottles labelled A and one labelled B from each place.

If milk has not kept we say that it has gone 'sour'. Sourness can be detected by taste, smell, and the fact that it 'curdles' when heated. However, it would not be a good idea to taste the milk. It is wiser to judge by smelling it and seeing whether it curdles or not.

7 Open the six bottles and smell each one.

8 Tip the milk into a labelled boiling tube and place the boiling tubes in a 600 cm^3 beaker of water.

9 Heat the beaker of water containing the tubes until it gently boils and look for evidence of curdling in the milk in the tubes.

10 Record the results in a table such as *table 17*.

Situation of bottles	Sample A				Sample B			
	After .... days		After .... days		After .... days		After .... days	
	smell	curdling	smell	curdling	smell	curdling	smell	curdling
refrigerator								
warm place								
out of doors								

Table 17
Table for recording results.

11 The following week, examine the remaining six bottles of milk in the same way and record the results.

Q3 From the results of your experiment, what conclusions are you able to form about improving the keeping qualities of milk?

Q4 How can you relate the results of your experiment to the treatment the milk received in the dairy?

Q5 From the results of your experiment, would you say that pasteurization or sterilization is the better method of treating milk?

Q6 So far we have only considered milk but what about other foods? How many different ways of preserving food are there?

You can collect information about this from advertisements and from the containers of preserved foods.

Q7 Try to explain why each method of preserving food you have mentioned does, in fact, keep the food from going bad.

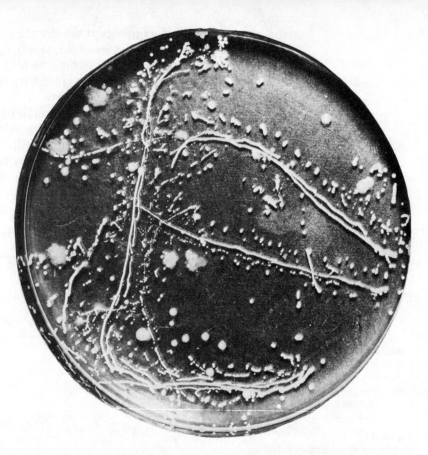

Figure 165
A culture of bacteria from the
tracks made by a fly.
*Photograph, Radio Times Hulton
Picture Library.*

Figure 165 shows the tracks of a fly written, so to speak, in
bacteria. A fly was allowed to crawl on a plate of nutrient
agar and the result was a chain of bacterial colonies.
Colonies of bacteria can grow in the same way on the
surface of meat if a fly walks over it after picking up
bacteria from somewhere else.

Q8 What does this suggest to you about the way to keep food
from being contaminated?

Eating contaminated food may result in food poisoning,
when micro-organisms act on the food to produce poisonous
substances which, when eaten, cause illness, or it may
result in food infection, when bacteria present in food are
eaten and multiply in our bodies, again causing illness.

Q9 In view of the information you now have about micro-
organisms and food,
a how should fresh food be stored before it is cooked or eaten?
b How should cooked 'left-overs' such as a joint of meat be
kept and used?
c How should frozen food be stored in shops and in the home?

8.3 Microbes and disease

Before going on to consider the prevention of disease and its cure, we must pause to consider what experimental evidence there is for the hypothesis that microbes do, in fact cause disease. It would be difficult, however, to perform a properly controlled experiment to show that a particular kind of bacterium does cause a particular disease. That would involve giving people diseases they would not have caught in the ordinary course of events. So we must study accounts of experiments done by other people.

Robert Koch, whose investigations into ways of growing bacteria have already been mentioned, was working on the connection between germs and disease. He followed a particular set of principles to establish whether any one micro-organism was the cause of a disease or not. These principles have become known as 'Koch's postulates' and they state:

1 The micro-organism supposed to be the cause of the disease must be found in large numbers in every case of the disease and in conditions which explain the symptoms and any pathological changes.

2 This micro-organism must be isolated in pure culture (that is, grown on agar).

3 The isolated micro-organism must reproduce the original disease when inoculated into other suitable healthy organisms.

4 It must be possible to re-isolate the same micro-organism from such infected organisms and grow it in pure culture.

Robert Koch applied his postulates to his own work on showing the connection between germs and disease. He successfully isolated and cultured a pure culture of anthrax *Bacilli* from the spleen of animals suffering from anthrax.

Louis Pasteur was also working on the same problem. In 1866, he was asked to investigate a disease that was killing off so many silkworms in France that the silk industry nearly came to an end.

A 'silkworm' is the caterpillar *(larva)* of a moth *(Bombyx mori)* which, when it forms a chrysalis *(pupa)*, spins around itself a delicate covering of long fine fibres to make a cocoon. To use the silk, the strands surrounding the cocoon are unwound and spun into silk thread.

Pasteur first examined under the microscope the ground-up bodies of silk moths which had died from the disease, and found bacteria, not present in healthy moths. Then he ground up eggs laid by infected moths. In these he also found bacteria of the same kind, which were not present in the eggs laid by healthy moths. He infected healthy moths with some bacteria and they quickly showed signs of disease. Those not infected remained healthy.

As you can guess, Pasteur's findings, although of great interest to scientists, were not at all popular with the merchants who sold silkworm eggs. Before these discoveries it had not mattered to them whether the eggs they sold were contaminated with disease, but their customers now became suspicious. In order to keep their stocks free from bacteria, they were forced to use more careful and expensive methods of culture.

Here, then, is an example of an actual attempt to show how, given the right conditions, an experiment can be devised to test a hypothesis.

Q1 Do you consider that Pasteur fulfilled Koch's postulates in his investigation to discover the cause of the disease in silkworms?

Q2 Would you now be able to suggest any improvements to Pasteur's investigations or do you think he had enough evidence?

Q3 Describe in detail how you would carry out an investigation to discover whether a microscopic fungus is responsible for the disease known as 'brown rot' in apples. A photograph of apples with brown rot is shown in *figure 166*.

Figure 166
Apples with brown rot at an advanced stage.
Photograph, R. J. Corbin.

The discovery that microbes are the cause of some diseases was a very important one. The following is a list of some of the diseases caused by the different types of microbes.

1 Diseases caused by bacteria

diphtheria	pneumonia
tuberculosis	tetanus
scarlet fever	syphilis
cholera	typhoid

Sore throats, abscesses, and food poisoning are also infections caused by bacteria.

2 Diseases caused by viruses

mumps	yellow fever
measles	influenza
chicken-pox	poliomyelitis
small-pox	myxomatosis in rabbits

Viruses are the cause of some diseases of plants. They are also the cause of the striped appearance of some kinds of tulip and the variegation in the leaves of some plants which are grown for this particular characteristic.

Figure 167 *(right)*
Variegation in the leaves of
Abutilon, another effect of a virus.
Photograph, R. J. Corbin.

Figure 168 *(below)*
A tulip flower, showing the striped
effect which is caused by a virus.
Photograph, The Royal Horticultural Society.

3 Diseases caused by fungi

In man, athlete's foot and ringworm are the result of fungus infections but many plant diseases are caused by fungi. For example, roses may have 'black spot', mildew, and rust, all of which are caused by different fungi.

Figure 169
a Rose leaves, showing black spot.
b Rose leaves, showing mildew.
Photographs, R. J. Corbin.

There are some 'diseases' which are not caused by microbes. A well known example is scurvy, which is a deficiency disease caused by a lack of vitamin C in the diet.

Introducing living things

It is not, however, enough just to know that microbes cause disease. It is also necessary to know how the disease-causing microbes are spread. There is more about this in the Background reading.

8.4 Can disease be prevented?

The work of Pasteur and others proved that microbes could cause disease. But the problem remained: could disease be prevented?

Figure 170
Statue of Edward Jenner.
Photograph, The Mansell Collection.

Nearly one hundred years previously an English country doctor, Edward Jenner (1749 to 1823), noticed that anyone who had suffered from the illness cowpox did not catch smallpox, which is a much more severe disease. Cowpox comes from handling cows and causes a rash of spots especially on the hands. It was quite a common thing in Jenner's time for girls who milked cows to catch cowpox and this probably gave him the idea. With a sterile needle he took a small amount of pus from the spots on the hand of an infected girl and transferred it to scratches made in the skin of an uninfected boy who soon got cowpox. The statue of Edward Jenner *(figure 170)* shows him performing this operation.

But Jenner's experiment was only half done. The boy recovered from cowpox and two months later, taking a grave risk, Jenner injected into the boy's arm pus from the spots of someone suffering from smallpox. Luckily for Jenner, the boy did not contract smallpox.

Our modern word 'vaccination' comes from the Latin *vacca*, which means a cow and reminds us that in Jenner's first vaccination against smallpox, he used the microbes of cowpox, a similar but much less severe disease. As the caricature in *figure 171* shows, the 'new inoculation' caused a good deal of suspicion and doubt in Jenner's time.

Figure 171
'The Cow-pock or the wonderful effects of the new inoculation.' From an etching by James Gillray in 1802, illustrating some of the absurd ideas held by people who were opposed to vaccination. *By courtesy of The Wellcome Trustees.*

The COW-POCK — or — the Wonderful Effects of the New Inoculation! — Vide the Publications of ý Anti Vaccine Society

Introducing living things

It is possible that a hundred years later, Jenner's discoveries gave Pasteur the idea of trying out similar operations on cattle. In France at that time many cattle were suffering from anthrax, a serious disease from which many of them died. Pasteur made a careful study of anthrax, and noticed that some animals developed the disease much more severely than others. He decided to inject two cows with a strong dose of anthrax bacteria, expecting them to die, but to his astonishment neither of the cows got the disease. Later, he found out that both animals had already suffered from anthrax. Could it be that they were now *immune* to it or, in other words, protected against it in some way? Pasteur argued that if it were possible to give an animal a mild attack of a disease, this might be sufficient to prevent it from getting the disease more severely later on.

One investigation led to another, and finally Pasteur succeeded in producing a weakened and harmless culture of anthrax bacteria. In 1881 he demonstrated in public the effect of vaccination in preventing anthrax. Pasteur injected thirty sheep with his anthrax vaccine. He then deliberately injected them with the anthrax germ, together with another group of sheep which had not had the vaccine. A few days later all the vaccinated animals were well but the unvaccinated ones were dead or dying.

Figure 172
Pasteur shows that sheep can be successfully vaccinated against anthrax.
By courtesy of the Wellcome Trustees.

Q1 What fact about the prevention of disease had Jenner and Pasteur established through their experiments?

Q2 In what way does Pasteur's experiment on anthrax differ from Jenner's experiment, as a scientific investigation?

The boy that Edward Jenner vaccinated with cowpox had been made immune to smallpox. Nowadays, most people are vaccinated against smallpox at least once during their lives. When Pasteur inoculated the cows with weakened anthrax bacteria, he had no real knowledge of *how* the vaccine worked. We now know that our bodies possess ways of attacking and repelling microbes; these are complicated processes which take place in the blood.

Imagine that there are two armies fighting on opposite sides. The enemy troops are the invading microbes which cause the disease, while the defending troops are cells and chemicals inside our bodies, able to fight the invaders and destroy them. First of all, the enemy has to gain entry into the body, through the nose or a cut in the skin, or in food. Once inside, enemy troops meet the defence troops, and provided these are there in sufficient numbers and are strong, they will immediately attack and kill the invaders. Constant battles of this kind are waged in our bodies without our knowledge. If the enemy troops manage to get to the battlefield under favourable conditions they will multiply. The defence troops have got little chance against so many and may lose the battle if reinforcements do not arrive.

Vaccination is like introducing a few enemy soldiers who encourage the defence army to multiply. Should the defenders now be faced with a large number of the enemy, they will be able to overcome them. Of course it is not really as simple as this, but the important thing to remember is that if we keep our bodies healthy, and therefore able to build up a good defence (immunity), there is much less chance that we shall become ill.

8.42 The effect of modern vaccines

Nowadays there are many kinds of vaccine which can be used to prevent a number of diseases. A range of vaccines is now available and these are given according to a schedule from the age of three months until the age of school-leaving. Vaccines included in the schedule are for diphtheria, tetanus, whooping cough, poliomyelitis, measles, and smallpox. A test is also given between the ages of ten and thirteen for tuberculosis. For people travelling abroad, vaccines are also available for yellow fever, cholera, and typhoid. Some countries will only admit travellers if they possess a current vaccination certificate for smallpox. This is a precaution to prevent such diseases being brought into the country from abroad.

Poliomyelitis has been a notifiable disease in this country since 1912. This means that the doctor must notify the Public Health Authority of every case of it which occurs. Thus, there are reliable records of the number of cases each year.

Information on the number of cases of poliomyelitis and the number of deaths resulting from it are given for selected years in *tables 18* and *19*.

Year	Number of deaths
1912	180
1932	178
1942	123
1947	688

Table 18
Number of deaths from poliomyelitis in England and Wales in selected years.
Data from the Office of Population Censuses and Surveys.

Year	Number of cases	Number of deaths in the year the disease was contracted	Number of later deaths from the same cause
1949	5982	643	8
1959	1028	66	21
1962	271	18	25
1963	51	3	34
1964	37	4	24
1965	91	3	15
1966	23	1	20
1967	19	1	14
1968	24	0	15
1969	10	0	10
1970	7	0	19

Table 19
The number of cases of poliomyelitis and the number of deaths from it in England and Wales in selected years.
Data from the Office of Population Censuses and Surveys and from the Department of Health and Social Security.

Draw a graph of the number of deaths from poliomyelitis from 1912 to 1970, using the figures given for selected years in *tables 18* and *19*, but excluding the 'late effects' given in the last column of *table 19*.

Q3 What was the pattern in the number of deaths from poliomyelitis up to the early 1940s?

Q4 What happened to the number of deaths in 1947?

Q5 Does the number of cases of poliomyelitis notified for the given years show a similar trend to the number of deaths (excluding 'late effects')?

Q6 Over which years did a noticeable change in the number of cases and number of deaths occur after 1947?

Q7 What do you suggest was the cause of this change?

Q8 What do the figures for the number of deaths from late effects of poliomyelitis tell you about this disease?

Q9 Do they show the same trend as the number of cases of poliomyelitis?

Q10 Can we now safely say that poliomyelitis has been eradicated from this country or are there still dangers of outbreaks such as occurred in the late 1940s?

Q11 What other diseases do you think might show a similar pattern to that of poliomyelitis in the number of cases occurring each year?

8.5 The cure of disease

There are many different ways of killing bacteria, and we have come across several of them already. But the main problem in curing a disease is not only to find a means of destroying the micro-organisms but also to avoid killing or injuring the patient. In the Middle Ages the only way known of treating illness was to use various extracts of plants. Some of these had limited value but the great majority were probably almost useless. Then, in 1910, the great German chemist Paul Ehrlich made a discovery which opened up a completely new field. He invented salvarsan – the first modern chemical to be used as a cure for disease.

You can make some observations for yourselves and try to arrive at some conclusions about one way in which some diseases can be cured. You will probably work in groups and each group will be provided with 3 Petri dishes labelled A, B, and C, each of which has been inoculated with a different type of bacterium.

Earlier, two discs were placed on the agar in each Petri dish and their positions were marked on the base of the dish with a P and an S. The Petri dishes have been in an incubator at 37°C for at least 18 hours.

1 Study each Petri dish in turn and record your observations by drawing circles the same size as the Petri dishes and showing what has happened in each case.

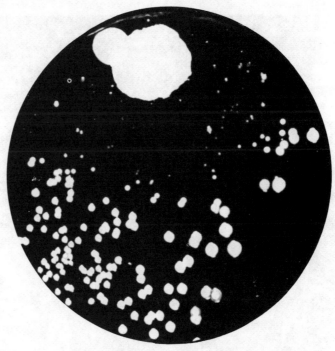

Figure 173
The Petri dish which Professor Alexander Fleming noticed in his laboratory in 1928.
Photograph, St Mary's Hospital, London.

2 Now look at the photograph in *figure 173*. Does this show a resemblance to any of your Petri dishes?

3 What do you think the discs contained and can you explain the effect they had on the growth of bacteria?

4 How do you explain the different results you get for the two discs?

8.51 The miracle drug

It often happens that scientists investigating one problem make an important discovery about something quite different.

This was what happened to Professor Alexander Fleming (later Sir Alexander Fleming) in 1928 when he was studying bacteria (*Staphylococci* – the kind of germs that cause boils and sore throats) in his laboratory at St Mary's Hospital, Paddington. He was examining some old bacterial plates (in Petri dishes similar to the ones prepared for you) when he realized that a mould (fungus) had grown on one of the cultures. There was nothing unusual about this. Moulds must have grown on cultures many thousands of times before but people just thought that the experiment had gone wrong. However, Fleming also noticed a very odd thing – the colonies of *Staphylococci* that had been growing in the area round the mould seemed to have disappeared. It looked as if some substance being produced by the fungus had destroyed the bacteria. So he cultured the mould, growing it in broth, and also carefully preserved the original plate, which was the one shown in *figure 173*.

Figure 174
Sir Alexander Fleming (1881 to 1955), in his laboratory.
Photograph, Keystone.

To his surprise he found that substances in the broth where the fungus was growing had the power of destroying a number of different bacteria, all kinds that cause human diseases. Further, it was effective even when diluted many times. The mould was later identified as *Penicillium notatum* and the mysterious substance which it had produced was named *penicillin*.

Fleming injected large doses of broth containing penicillin into mice and showed that it was not harmful. But while doing these experiments he also found that the broth lost its power of killing bacteria if kept for any length of time. Penicillin seemed to be a substance which was difficult to preserve. Perhaps it decomposed when kept in laboratory conditions.

Introducing living things

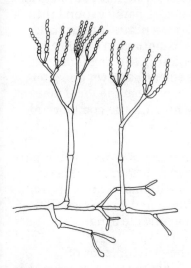

Figure 175
Penicillium mould (× 300). The mould grows chains of spores which can be carried in the air or the soil and can grow to form another mould.

Curiously enough there was little interest in the first reports on penicillin. Scientists were busy thinking of other ways of curing human disease. Then, in 1938, Professor Howard Florey (the late Lord Florey) and Dr Ernst Chain (now Professor Chain) started work at Oxford on a study of substances produced by living things that help to kill the harmful bacteria which attack them. On reading through the literature, Chain came across Fleming's account of penicillin, and decided to repeat his experiments. Chain found exactly the same things as Fleming, including the difficulty of preserving penicillin in the laboratory. Now Chain took the experiments a step further and injected different kinds of harmful bacteria (e.g. *Staphylococci* and *Streptococci*) into separate groups of mice. Each group was divided in two. One-half also received a dose of penicillin broth, the other (the control) did not. What would happen to the diseased animals? It was an exciting moment and the scientists stayed up all night in the hope of finding the answer. The experiment was a complete success – the untreated mice died, while those that had received penicillin survived. But the problem remained of isolating and preserving this new 'miracle drug'.

It was now June 1940 and wartime. England was awaiting invasion. Fearing that Fleming's precious cultures of *Penicillium* might be lost if the Germans landed, the scientists soaked the linings of their coats with broth containing the mould, so that as long as one of them survived the culture could be grown again. (The spores of many fungi can remain alive in a dried state for long periods of time.)

The first patient on whom it was tried was a policeman ill with blood poisoning in an Oxford hospital. On being given penicillin he showed a remarkable improvement but unfortunately supplies ran out and he died. Then in April 1941 it was used again for a boy with an infected hip wound and on a patient with an enormous carbuncle. In both cases treatment was successful. But Florey realized that in spite of all the hard work at Oxford it was going to take many months to obtain pure penicillin even in small quantities, owing to the complex processes needed to extract it. Since it was wartime, there was an urgent need for drugs to cure infected wounds and other diseases. So, in 1941, Florey flew to the United States, taking supplies of the cultures with him. There he joined a group of American scientists working at the Northern Regional Research Laboratory, Illinois.

The search for a way of isolating penicillin now started on a big scale. Soon it was discovered that the yield of the new drug could be increased by twenty times if the *Penicillium* was grown in corn-steep liquor (the liquid left behind after maize had been soaked in water before being ground into flour). Meanwhile many other strains of the fungus from different parts of the world were examined to see if they would give larger amounts of penicillin. Strangely enough, one of the best was found on the scientists' own doorstep, growing on a mouldy melon obtained from the local market. This was *P. chrysogenum* from which all the commercial strains used today were derived.

Eventually a way was found of making and preserving pure penicillin and several manufacturers began to make it commercially. By this time, trials with many animals and different kinds of dangerous bacteria had confirmed the value of penicillin. It was indeed a miracle drug.

For the part they had played in the development of penicillin, Fleming, Florey, and Chain together received the Nobel Prize for medicine in 1945. Since then numerous other

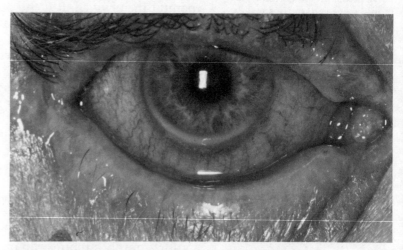

Figure 176
a An eye infection before treatment.

b The eye has been treated with an antibiotic.
Photographs, The Institute of Ophthalmology, University of London.

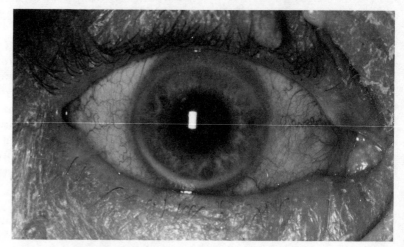

Introducing living things

substances similar to penicillin have been discovered. One of the best known is streptomycin which has proved successful in the cure of tuberculosis. All are produced by different kinds of micro-organisms and act by killing other living organisms, particularly bacteria, which come in contact with them. These substances are called *antibiotics* (from the Greek, *anti* = against; *bios* = life), indicating that certain kinds of life cannot exist when they are present.

Today penicillin is used by doctors and in hospitals all over the world and has been responsible for the great advances made in the cure of blood poisoning, venereal diseases, pneumonia, and meningitis.

8.6 Tracking down the source of infection

In 1964 there was a serious outbreak of typhoid in Aberdeen. The epidemic reached over 400 cases before it was controlled. Study the following account of the progress of the epidemic and the measures taken to control it, and then try to answer the questions.

9 May The city hospital in Aberdeen receives 4 patients with an undiagnosed illness.

20 May The illness is diagnosed and confirmed as typhoid. Two possible sources of infection – ice cream and cold meat – are traced.

21 May, morning Ice cream is ruled out as a source of infection. Possible sources are narrowed down to three. *Afternoon* Suspected source of infection is narrowed down to one shop.

22 May Source of infection for all cases is narrowed down to corned beef from one large tin. Corned beef is accepted as a source of infection by the Health Department in Aberdeen. All general practitioners in the town are informed of the source of infection. 4 health education education officers, 72 health visitors, and 10 sanitary inspectors are working on the epidemic, under the direction of the Medical Officer of Health.

23 May There are now 48 confirmed cases of typhoid in the city hospitals. A public health laboratory in London identifies the bacterium responsible for this outbreak of typhoid. It is a type found only in Spain, Latin America, and the southern United States. The bacteria were contained in a large tin of corned beef imported from Argentina.

25 May Extra cases of typhoid occur where the patients have not eaten corned beef. Investigation now shows other cold sliced meat was infected.

---

28 May It is estimated that 10 000 people had bought meat from the suspect shop between 6 and 23 May. Some of them will be developing typhoid.

29 May Within three hours 29 more cases are admitted to hospitals.

1 June There are 64 new cases of typhoid admitted to hospitals. Aberdeen Medical Officer closes schools, cinemas, dance halls, bingo halls. Television and radio statements are made about the danger of pre-cooked food and the necessity to wash the hands after using the lavatory.

2 June The Government withdraws from sale in shops all corned beef tinned by suspect firms.

3 June There are 30 new cases of typhoid all from the same source.

5 June There are 17 new cases of typhoid.

8 June There are 10 new cases of typhoid.
£400 worth of meat from a butcher's shop in Aberdeen is destroyed and the shop is closed.

9 June There are 4 new cases of typhoid.
Another shop in the city is closed – this time a fruit shop.

10 June There are 15 new cases of typhoid.

17 June The Medical Officer of Health for Aberdeen reports that the outbreak is controlled and the danger over.

18 June No new cases of typhoid have been admitted to hospitals; the restrictions are lifted.

Q1 When the illness had been identified as typhoid, what questions would have been asked by the Health Department's team of workers?

Q2 Why do you think the source of infection was narrowed down to ice cream and cold meat and why was the ice cream eventually ruled out?

Q3 How did the shop come to sell the corned beef or people to eat it if it was contaminated with the bacteria which caused typhoid?

Q4 Suggest possible ways in which other sliced cold meat could have been infected.

Q5 Can you give any special reasons why the Medical Officer of Health decided to close schools, cinemas, dance halls, and bingo halls?

Q6 Why did television and radio statements refer to the danger of pre-cooked food and to the need to wash the hands after using the lavatory?

Q7   Why were the two shops closed on 8 and 9 June?

Q8   What would have led the Medical Officer of Health to report on 17 June that the outbreak of typhoid was controlled and the danger over?

This was a splendid piece of detective work and the investigators found that the beef was infected in the first place because the South American factory did not use chlorinated water during the canning process.

Background reading

Plagues, public health, and people

In the first century A.D. the average length of life was about twenty-two years. By the nineteenth century the expectation of life in western countries had increased to forty-three years for women and to forty-one years for men. The expectation of life for a woman in this country now is seventy-five years and for a man seventy years.

There has also been an increase in the size of the population. During the nineteenth century the population of Great Britain more than trebled. In 1801 the population was 11.6 million and in 1900 it was 38.2 million. In 1972 it was 49.3 million.

Why can we expect to live longer than our great-grandparents and our other more distant ancestors did? There is more than one explanation for these happenings and Chapters 7 and 8 describe some of the factors which have ensured that men live longer and that the population grows rapidly.

In this country today we do not have outbreaks of diseases such as cholera and bubonic plague, nor are we afraid of catching diphtheria or tuberculosis. We take for granted that the water from our taps is safe to drink and that our sewage and other household waste is disposed of safely.

As recently as the last century living conditions were very different. People living in the nineteenth century may have thought that they were much more fortunate than people living before them. For instance, the Black Death, or bubonic plague, which was brought to England in 1348 and killed a large part of the population, was no longer a serious threat. The last serious outbreak had been the Great Plague in 1665.

It is, however, as well to remember that bubonic plague still exists in parts of Africa and Asia. In 1965 2000 cases of bubonic plague were diagnosed in Vietnam. We now know that it is passed on to man by the rat flea, so it is more easily controlled but the chance of recovery from it is still rather poor.

A disease which was still a serious threat in the nineteenth century was cholera. This was brought into the country in 1831 through the port of Sunderland and it quickly spread throughout the country. It was particularly serious in the larger towns.

Cholera is a most unpleasant disease, being a particularly violent form of diarrhoea. In the nineteenth century there was no effective means of treatment and the death rate was very high. A look at the living conditions in many towns at that time helps us to see why cholera spread so rapidly.

'At such times, a stranger, looking from one of the wooden bridges thrown across it at Mill Lane, will see the inhabitants of the houses on either side lowering from their back doors and windows, buckets, pails, domestic utensils of all kinds in which to haul the water up; and when his eye is turned from these operations to the houses themselves, his utmost astonishment will be excited by the scene before him. Crazy wooden galleries common to the backs of half-a-dozen houses, with holes from which to look out upon the slime beneath; windows, broken and patched, with poles thrust out, on which to dry the linen that is never there; rooms so small, so filthy, so confined, that the air would seem too tainted even for the dirt and squalor which they shelter; wooden chambers thrusting themselves out above the mud, and threatening to fall into it – as some have done; dirt-besmeared walls and decaying foundations; every repulsive lineament of poverty, every loathsome indication of filth, rot, and garbage; all of these ornament the banks of Folly Ditch.'

That was Dickens's description, at the end of *Oliver Twist*, of how the inhabitants of houses near Jacob's Island on the Thames got their water from Folly Ditch.

It was not only in London that such conditions were found. All the large cities in the country had similar problems. The courtyards and alleys that existed then made the problem worse. With no proper sanitation in the houses, human excrement, together with other household rubbish, was kept in the houses until people could stand it no longer and then it was thrown outside where a mound accumulated. Some of this would eventually drain into the nearby ditches or rivers which were the only water supply so that people would have to use this contaminated water for washing and drinking.

Figure 177
Jacob's Island, Bermondsey, about
1810. The stream was the sole
water-supply *and* means of
disposing of sewage for the people
living in the houses there.
*By courtesy of the Wellcome
Trustees.*

Only a few of the main streets were swept and so piles of
rubbish accumulated everywhere.

Figure 178
An engraving from the painting
'Night' by Hogarth, in his series of
paintings 'Times of the Day'.
The Mansell Collection.

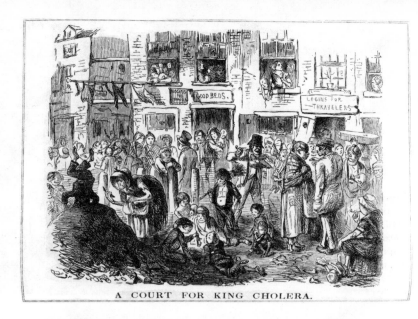

A COURT FOR KING CHOLERA.

Figure 179
A drawing of one of London's courtyards, published in 1852. Notice the children playing in a mound of excrement and rubbish. A woman is scavenging in it for food, among the rats. Presently, the mound will pollute the water supply and spread cholera. *The Mansell Collection, reproduced by permission of* Punch.

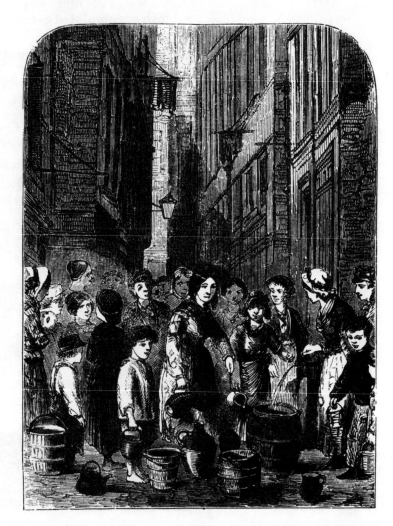

Figure 180
Fryingpan Alley, Clerkenwell, in 1864. The one stand-pipe which you can see was the sole water supply for many houses and it was turned on for only twenty minutes a day. *By courtesy of The Wellcome Trustees.*

Thus, well into the nineteenth century there was no regular collection of rubbish and water safe to drink was virtually unobtainable in London. A few rich people had a piped supply. Others had to collect rain water or rely on a well-supply, often contaminated by the filthy water draining into it from the street, or use public pumps situated in the street. Water sellers toured the streets and stealing water was a common crime.

Many people preferred to drink alcohol, mostly beer and gin, with the result that drunkenness was another of the social problems of the time.

The cartoonists of the day often drew attention to the unfit state of the water for drinking as you can see from *figure 181*.

A DROP OF LONDON WATER.

Figure 181
'A drop of London water.'
The Mansell Collection, reproduced by permission of Punch.

Gradually people realized the need to clean up the streets, to improve the water supply, and to have a sewage system. Thanks to the efforts of public-spirited men, who drew attention to the conditions, the first improvements came about.

One of the men who achieved a great step forward was Edwin Chadwick (1800 to 1890).

Figure 182
Edwin Chadwick. This print was published in 1848, the year the Public Health Act was passed, largely as a result of his work. *Ronan Picture Library*.

Chadwick was a civil servant who had trained as a barrister and journalist. He, with others to help him, carried out a survey in the poor districts of London and in 1842 the findings were published in 'The report on the sanitary conditions of the labouring population'. He described the appalling conditions which existed and set forward drastic measures for changing them. This report, which became known as Chadwick's Report, is possibly still the most important document in the history of public health.

Unfortunately, the government of the day was not prepared to act on his recommendations immediately, and it was 1848 before the government eventually passed the first Public Health Act.

This is perhaps strange as the Houses of Parliament were well known for the 'bad air' which came from the River Thames. Members of Parliament could not walk out on the balcony in the summer months without holding handkerchiefs across their noses and mouths. In 1858 the work of Parliament was nearly brought to a standstill.

The sewage of nearly three million people had been allowed to accumulate in a vast cesspit in the middle of the city. The hot sun shining on this produced a smell which became known as the Great Stench of London.

The Law Courts did close but the Houses of Parliament remained open although blankets soaked in chloride of lime were hung at the windows and disinfectants were used in all the rooms. Much to many people's amazement there was no serious outbreak of disease that summer.

Ten years before the Great Stench of London, in 1848–9 another epidemic of cholera swept the country and at the same time a pamphlet was published by a Dr John Snow (1813 to 1858). It was called 'On the mode of communication of cholera'.

Figure 183
Dr John Snow in 1856.
Photograph by courtesy of The Wellcome Trustees.

In his pamphlet, John Snow put forward the theory that cholera was spread as a result of drinking water which was contaminated with sewage from the houses of cholera victims.

In 1854 he had the opportunity to test his theory as cholera broke out once again. This time Snow made a study of all the cases in the Soho area of London and he discovered that almost all the victims had drunk water from the public water pump in Broad Street. Dr Snow suggested to the authorities that the pump handle should be removed. This was done and people were forced to go elsewhere to obtain their water. The epidemic was stopped and it was later shown that the water in the well supplying the Broad Street pump was contaminated by a cesspool serving houses in which a cholera patient had lived.

It was not only well water that was unfit to drink. Piped water which was gradually becoming available could also be dangerous especially if the original source was contaminated with sewage.

Figure 184
'Source of the Southwark water works.' A drawing by George Cruikshank from the Cribb Collection.
By courtesy of the Royal Institute of Chemistry.

Introducing living things

There was some progress and then, in 1875, a new Public Health Act forced local authorities to provide adequate drainage and sewage systems in their towns and a safe water supply. Eventually public health authorities got rid of water-borne diseases by chlorinating the water supplies.

Nowadays, although many coastal towns discharge raw sewage into the sea, inland towns have sewage works where the sewage is treated, so that uncontaminated water is eventually returned to rivers or lakes.

Today The water supplied to our houses is very unlikely to contain any harmful bacteria and we accept that it is quite safe to drink. We also take proper sanitation and drainage for granted. Do not forget, though, that not many years ago things were very different and it is due to the efforts of men like Edwin Chadwick and John Snow that efficient services supplied by the Local Council first came about.

So, because of improvements in hygiene, because we know how disease is spread, and because we now have vaccines, epidemics of killer diseases such as typhoid, smallpox, diphtheria, and cholera are no longer a threat in the western world. There are countries in other parts with low standards of hygiene and sanitation where cholera and typhoid are still present dangers and we know that they could be a thing of the past, as with the Black Death in this country.

Are epidemics of diseases to be feared today in this country? From time to time newspapers report that widespread outbreaks of influenza are forecast for the winter months. Influenza is no new disease but it has never had quite the same publicity as smallpox or cholera.

With all these diseases which seem to be a thing of the past or no longer to be feared, it is well to remember that they are still with us and if we are careless about using the measures for keeping them under control it is likely that they will strike again. The outbreak of poliomyelitis in autumn 1961 in Hull was a reminder of this fact. But because of the efficient actions of the Local Health Authority this outbreak was controlled very quickly. Ample supplies of the oral vaccine were readily available and as a result of an advertising campaign in newspapers, on radio and television, and in posters displayed in shops, most of the people living in Hull had been vaccinated in just over two weeks from the time poliomyelitis was discovered there.

Because we have an efficient and up-to-date Health Service, this country is often able to send vaccines and other medical supplies as well as doctors and nurses to other countries when disaster hits them.

There is, however, one black spot in this success story. Although most diseases have shown a rapid decrease in numbers there is one group of diseases of which the number of cases is rapidly increasing, with something like one thousand new cases a week. Does this startle you? They might well be called very dangerous diseases. They are the venereal diseases, known as VD for short.

There are two common venereal diseases, one called syphilis which is caused by a spiral micro-organism called *Treponema* and gonorrhea caused by a bacterium called a gonococcus. Both of these diseases can be cured with penicillin and the way in which they are spread is known. Why then are there 50000 people suffering from gonorrhea in this country today and 3000 suffering from syphilis? And why is the number increasing especially in young people under twenty-one?

Both of these diseases are passed from person to person during sexual intercourse and the people most likely to get it are those who move from partner to partner. Men and women who have sexual intercourse with just one person (their wife or husband) are not likely to run the risk of ever getting VD.

There are clinics for diagnosing and treating VD but this is a sad example of where advances in medicine and an efficient Health Service have failed. Why have they failed? Whether people catch the disease or not depends on how they behave and only they can protect themselves by behaving in a way which will not put their health at risk. The control of venereal diseases is not dependent on advances in medicine and good living conditions and it is up to individual people to make their control a success story instead of the modern tragedy it is.

Introducing living things

Finding out about insects

9.1 Observing insects

Insects form the largest group of animals and the number of different types (species) of insects is greater than the number of species of all other animals put together (see *figure 206)*. So far, about three-quarters of a million species of insects have been named, but many more are being discovered each year. With the exception of the sea, insects can be found in every habitat from the Poles to the Equator and so the opportunities for observing them are almost limitless. Many insects are easy to keep in captivity and it is difficult to decide which kinds to choose for studying in the laboratory. Probably it will be best to start with large insects with interesting life histories. Grasshoppers are large and easy to keep but our native ones are difficult to breed in captivity. But the locust, which is a large kind of grasshopper not found in this country, breeds well in the laboratory.

The stick insect is another animal which is easy to keep and breed and its life history differs from that of a locust in an interesting way.

9.11 Keeping insects in the laboratory

Locusts
There are several different kinds of locusts, but the best species for keeping in school is the African migratory locust *(Locusta migratoria migratorioides)*. This is found in the tropical regions of Africa as well as in European countries bordering the Mediterranean.

Study the way locusts are kept in your laboratory and, as most cages have the same basic features, try to answer these questions.

Q1 What is the purpose of the false floor of the cage?

Q2 What is the purpose of the hole covered with perforated zinc in the roof of the cage?

Q3 Why is there a 'door' in the roof?

Q4 Why are there two electric lamp bulbs in the cage and why should one of them be switched off at night?

Q5 What is the purpose of the twigs (or netting)?

Q6 What are the sand-filled tubes for at the bottom of the cage?

Q7 Why is it that when you give the locusts fresh grass (or other fresh vegetable food) there is no need to provide any water, whereas if you feed them on bran you must provide water?

Q8 What routine attention do you consider you should give to the locusts each day?

Stick insects
Stick insects *(Carausius morosus)* do not occur naturally in this country but are found in eastern countries. They can easily be kept in an insect cage or in a glass tank or aquarium with a perforated lid.

Although they live in warm regions they will survive well at normal laboratory temperatures. They can be fed on privet leaves which should be placed in a jar of water in the cage and changed once or twice a week. In winter, if privet is not readily available, you may persuade them to feed on ivy. When you change the privet, remove the droppings from the cage, and collect any eggs. Keep the eggs in a small dish or matchbox without a lid, in the cage.

9.2 What are the special features of insects?

By examining a locust, a stick insect, or any other insect in greater detail, you should be able to discover some of the features which are typical of insects.

9.21 Examining a locust

1 Study a locust and, with the help of *figure 185*, find the different parts of its body – its head, thorax, abdomen, and wings.
2 Now look more carefully at the way in which the wings and legs are joined onto the body.

Q1 To which part of the body are
a the wings attached
b the legs attached?

Q2 What do you notice about the structure of the third pair of legs?

Figure 185
A photograph and diagram of the
male African migratory locust,
Locusta migratoria migratorioides.
a *Photograph by Glenda
Colquhoun/The Centre for
Overseas Pest Research.*

3rd leg

forewing covering hind wing

thorax

1st
segment

2nd
segment

3rd
segment

antenna

head

mouth parts

abdomen

1st leg

2nd leg

b *Diagram after Thomas, J. G.
(1963) Dissection of the locust,
Witherby.*

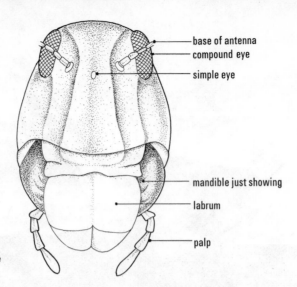

base of antenna
compound eye
simple eye

mandible just showing

labrum

palp

Figure 186a
The head of a locust, seen from the front.

Figure 186a shows an enlarged drawing of the head of the locust seen from the front.

3 Try to look at your locust from the same angle.

Q3 What can you see about the position, shape, and structure of the eyes?

4 Still looking at the head from the front, find the upper lip or *labrum*. If you could lift the labrum up with a blunt seeker you would find a pair of strong jaws or *mandibles* beneath. *Figure 186b* shows the mandibles in position. Notice also the jointed palps visible on either side of the labrum.

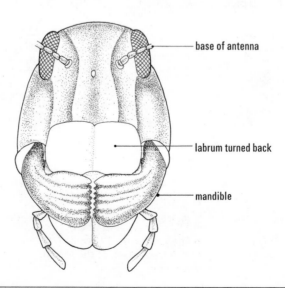

base of antenna

labrum turned back

mandible

Figure 186b
The head of a locust, with the labrum turned back to show the mandibles. The cutting edges of each mandible meet in the centre.

5   *Figure 187* shows one of the hind legs enlarged. On your
    specimen find the various parts of the leg – *femur*, *tibia*, and
    tarsus.

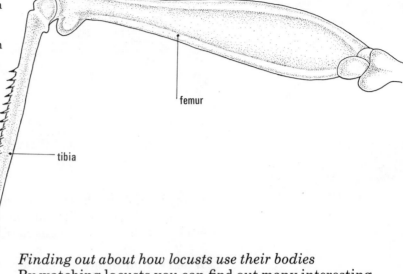

Figure 187
The third right leg of a locust, seen
from the outside. The joints are
flexible and allow each part of the
leg to bend. The tibia is armed with
a row of strong spines.

femur

tibia

tarsus

Finding out about how locusts use their bodies
By watching locusts you can find out many interesting
things about their habits and how they use their bodies –
for example, how they feed and move.

Watch locusts and as you do so try to answer the following
questions.

Q4   How does the locust grasp a blade of grass while it feeds?

Q5   How do the labrum and jaws move as it is feeding?

Q6   Are any other parts of the body moving while it is feeding?
     You will probably be able to see that much of the body of the
     locust is covered by a hard jointed skeleton.

Q7   Can you see the reason for the thin membranes between the
     joints?

Q8   If any locusts are walking up the sides of the cage, can you
     see how they cling onto the cage?

9.22 Examining a stick insect

1   With the help of *figure 188*, identify the different parts of the
    stick insect's body – its head, thorax, abdomen, three pairs
    of legs, and antennae.
2   Look at one of the legs.

Q9   Is this leg built on the same plan as those of the locust?

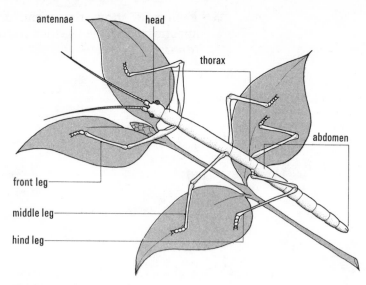

antennae   head

thorax

abdomen

front leg

middle leg

hind leg

Figure 188
An adult stick insect, *Carausius morosus* (× 0.8).

Using your own observations on the living insects and on photographs, make lists of the ways in which locusts and stick insects *a* resemble each other, *b* are different.

When watching stick insects you may not see much activity as they are nocturnal (that is, they are most active at night).

9.3 Investigating the life histories of insects

9.31 The life history of the locust

Mating and egg-laying
In a cage of adult locusts, if there are males and females they will often be seen together in pairs, the male on top of

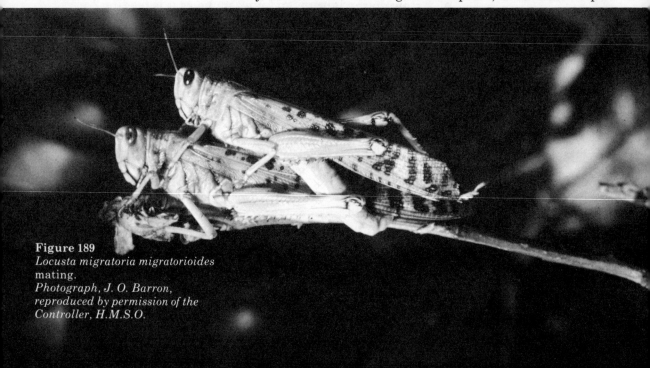

Figure 189
Locusta migratoria migratorioides
mating.
*Photograph, J. O. Barron,
reproduced by permission of the
Controller, H.M.S.O.*

the female. When they are in this position they are mating and may stay together for several hours.

1   If possible, look closely at a pair of locusts mating. Also see *figure 189*.

Q1   What differences do you observe in their appearance?

Q2   What do you observe about their mating behaviour and what do you suppose is taking place?

In the desert regions of Africa, where locusts are usually found, the females lay their eggs in the warm, damp sand and after the rainy season.

Q3   How do we copy these conditions in the laboratory?

Sooner or later you should see one of the females sitting on top of one of the tubes, often with a male locust still on her back. Watch carefully and try to make out what she is doing. You may be able to see the female's abdomen pushing down through the sand. The valves at the end of the abdomen (see *figure 190*) are opening and closing and, as they do so, they are pressing the particles of sand apart. In this way the whole abdomen is pushed further and further through the sand, getting longer and longer as it does so, until it very nearly reaches the bottom of the tube. Then the female will start to withdraw the abdomen and as she does so, she fills the hole in the sand with a frothy substance in which she lays her eggs.

Figure 190
The end of the abdomen of a female locust pushing through sand.

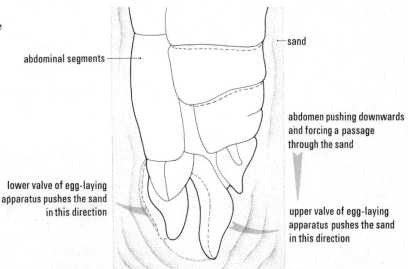

abdominal segments

sand

abdomen pushing downwards and forcing a passage through the sand

lower valve of egg-laying apparatus pushes the sand in this direction

upper valve of egg-laying apparatus pushes the sand in this direction

By the time the female locust has almost withdrawn her abdomen, the column of froth will be packed with eggs. The froth hardens round the eggs and the whole structure is called an egg pod.

Examining an egg pod
If an egg pod has been laid against the side of the tube you will be able to see it through the glass. If you cannot see anything, carefully scrape away the loose sand at the top of the glass tube. If an egg pod has been laid, the top of it will show as a hard lump in the sand.

You can easily examine an egg pod and find out more about the eggs inside it.
1 Empty out the sand from one of the glass tubes and take out the egg pod made by the locust.
2 Measure and record its length in millimetres.
3 With a pair of sharp pointed scissors, carefully slit the egg pod up one side without damaging the eggs inside.
4 Draw what you see.
5 With a blunt seeker, carefully separate the eggs from the sand and froth and count them.
6 With a hand lens, examine one of the eggs and measure its length in millimetres. *Figure 191* shows you what you should look for.

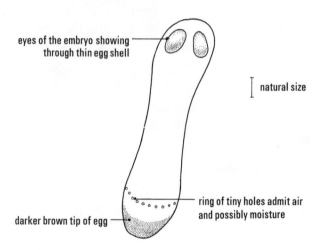

eyes of the embryo showing through thin egg shell

natural size

ring of tiny holes admit air and possibly moisture

darker brown tip of egg

Figure 191
The egg of a locust.

Q4 Can you think of an advantage that comes from laying the eggs in damp sand?

Q5 Can you think of any advantages in the female surrounding her eggs with froth?

Q6 You saw that the female starts to lay her eggs at the bottom of the tube. What is the advantage of this?

Introducing living things

Q7   How are the eggs arranged in the egg tube?

Does temperature affect the length of incubation of the eggs?
Under natural conditions in the desert, the temperature of
the sand in which the female locusts lay their eggs is usually
about 34 °C. At this temperature the eggs take eleven days
to hatch. You could try keeping one egg pod at 34 °C, and
one pod at a lower temperature. In this way you could find
out if eggs kept at a lower temperature take longer to
hatch.

1   Use two of the glass tubes in which there are freshly laid
     egg pods. Label one A and the other B and add to the
     labels the date when the pods were laid.
2   Add 5 cm³ distilled water by means of a dropper or syringe
     to each tube and put it in a small polythene bag. Tie the
     neck of the bag and make a few small holes in it. The bag
     keeps the sand moist and the holes allow air in.
3   Leave tube A in the big locust cage at 34 °C. Put B in an
     incubator at 28 °C.

Q8   What is the purpose of tube A?

4   You will expect the eggs in tube A to hatch about eleven
     days after they were laid. So you must remember to watch
     both tubes for signs of the young locusts hatching, from the
     eleventh day onwards.
5   Make a note of the date on which the young locusts hatch
     in tubes A and B.

Q9   Can you give a reason for any difference in the time it took
      the eggs to hatch?

Once the eggs start hatching, remove the tube from its bag
and put it in a cage.

What does an embryo locust look like?
In the course of her life a female locust will make several
egg pods, each about the size of a grain of rice and each
containing anything between 30 and 100 eggs. Inside each
fertile egg a young locust nymph is developing.

Look at the specially prepared egg under the microscope.
Can you make out the head, abdomen, and legs of the
future locust nymph?

The young locusts hatch and develop
If you are in the laboratory at the right time you may see
the young locusts actually coming up out of the sand in the
egg tubes. They are a pale colour and will probably stay on
top of the tube for a while. During this time they become
darker in colour and soon start to hop about.

At this stage, they are called hoppers. They will soon begin feeding. Thus you should provide grass and a dish of bran in each of the cages. They will also need a few twigs to climb.

You have found that the locust, like almost all other insects, is covered with a hard skeleton which will not stretch, so when the hoppers feed they cannot get any bigger unless they change their skins. If you watch the hoppers carefully, you may see one or two of them climbing up the twigs in the cage and grasping a twig with their feet. As they hang head downwards the skin splits along the back of the hopper and a new nymph emerges. This skin splitting or moulting is called *ecdysis*.

At first, the newly emerged nymph is pale in colour and quite soft, but the skin later hardens and becomes darker. The nymphs will soon start to feed again and, of course, to grow so that once more they must go through an ecdysis. Each nymph stage between one ecdysis and the next is called an *instar*. When the hoppers are newly hatched they are called first instars. In the locust, the nymphs pass through five instars before they finally emerge as adults (see *figure 192*).

If you watch the cage of hoppers daily you will be able to record when they moult and become second instars and so on. However, not all the nymphs will moult at the same time so the best thing to do is to record the date on which about half the nymphs have changed to the next instar. Put your results in a table on the lines of *table 20*.

Date when eggs hatched . . . . . .   Temperature of cage . . . . . .

	Date when half the nymphs have changed to the new stage	Notes on the appearance of the nymphs
2nd instars emerged 3rd ,, ,, 4th ,, ,, 5th ,, ,, Adults emerged		

Table 20
Record of hatching and developing of locust.

If two cages are available you can lower the temperature in one by using bulbs with a smaller wattage. If you then place equal numbers of hoppers in each cage and treat both sets in an identical way you can compare the rates of development at different temperatures.

Introducing living things

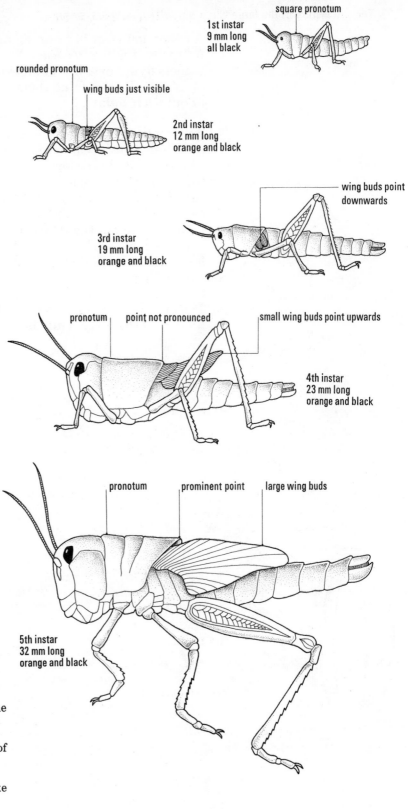

1st instar
9 mm long
all black

square pronotum

rounded pronotum

wing buds just visible

2nd instar
12 mm long
orange and black

wing buds point
downwards

3rd instar
19 mm long
orange and black

pronotum

point not pronounced

small wing buds point upwards

4th instar
23 mm long
orange and black

pronotum

prominent point

large wing buds

5th instar
32 mm long
orange and black

Figure 192
The development of the locust. The
labelling indicates the characteristics you should look for to
distinguish the nymphs. The size of
a nymph is not always a reliable
guide.
Note also that the large saddle-like
covering of the first thoracic
segment is called the pronotum.

9.32 The life history of the large white butterfly *(Pieris brassicae)*

A common insect in a garden in summer is the large white butterfly *(Pieris brassicae)*. You can usually see these insects flying over a vegetable patch and sometimes settling on plants. By watching them we can learn quite a lot about their habits.

As they fly around you may see two butterflies coming together. They seem to touch each other and then separate. They may do this several times and even settle on a plant together. They are probably pairing and soon afterwards the female may start to lay a batch of eggs. She lays her eggs in batches of 30 to 100 and you may be able to find them in patches about the size of a 1p piece on the underside of cabbage leaves. They will be bright yellow. If you look at a large number of cabbage leaves you may also find caterpillars.

Figure 193
Stages in the life history of the large white butterfly, *Pieris brassicae.*

The photographs in *figure 193* show the stages in the life history of the large white butterfly.

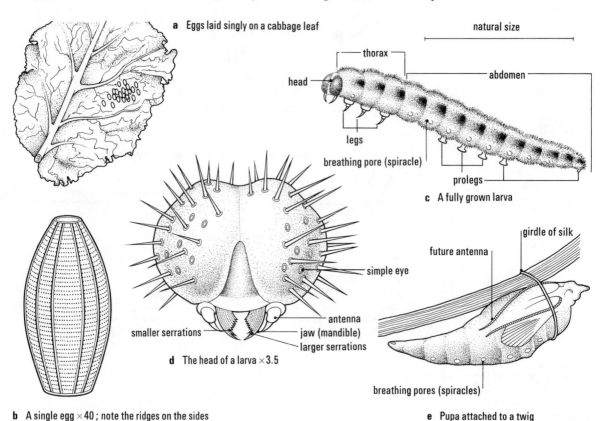

a Eggs laid singly on a cabbage leaf

natural size

head · thorax · abdomen
legs
breathing pore (spiracle)
prolegs

c A fully grown larva

simple eye
antenna
smaller serrations · jaw (mandible)
larger serrations

d The head of a larva ×3.5

girdle of silk
future antenna

breathing pores (spiracles)

e Pupa attached to a twig

b A single egg ×40; note the ridges on the sides

Investigating the life history of the large white butterfly
If you can obtain some eggs you can hatch them and observe the caterpillars develop. You can also compare the life history of this insect with that of the locust.

This is what you should do:

1   Examine the egg with a hand lens and draw it. Note the colour. Compare the egg you have drawn with the one in *figure 193b*.
2   Prepare a small plastic box, about $13.0 \times 7.5 \times 2.0$ cm, by lining it with a sheet of absorbent paper (blotting paper or even newspaper will do) cut just larger than the bottom of the box. Make a small cut from each corner towards the centre. Press the paper into the box and it will then fit the bottom and up the sides.
3   Put the piece of cabbage leaf on which eggs have been laid into the box and add another small piece of fresh cabbage leaf, putting the lid on firmly.
4   The eggs will hatch more quickly if they are kept warm at about 24°C but they should not be put over hot pipes or in the sun.
5   Every day examine the eggs through the transparent lid to see how they are getting on. Are they still yellow? If they are becoming darker it means that they will soon hatch, since the black colour is due to the head of the caterpillar inside the egg shell.
6   Note the date on which the caterpillars hatch from the eggs and record it as in *table 21*. What do the caterpillars do as soon as they hatch? It may be possible for you to measure a caterpillar (in mm) although it will be very small. If you succeed, record its length in your table of growth.

		Notes of changes in appearance of caterpillars
Date eggs hatch		
Length (in mm) of caterpillars at hatching		
Length at 1 week old		
Length at 2 weeks old		
Length at 3 weeks old		

Table 21
Table for recording details of the growth and appearance of a caterpillar.

7   The caterpillars will need to be fed and to do this you should first remove the paper from the box with the cabbage leaf and caterpillars on it, then re-line the box with a fresh piece of paper.

Next, without touching the caterpillars with your fingers, either brush them with a paint brush back into the box or, if they are on the old cabbage leaf, leave them where they are, but put a fresh leaf in the box beside them. Replace the lid. They can stay in the small box until they are about 10 mm long, and they must be transferred to a larger plastic box (about $15 \times 12 \times 6$ cm). You should change the paper and give them fresh food every day, otherwise they may get diseased.

8   During each lesson, measure one of the caterpillars. When they are about 30 mm long you should draw one, labelling the parts of the body.

Q10   Has their colour changed while they have been growing?

Q11   How does a caterpillar use its legs and prolegs in movement?

You will remember that in the locust the skin was hard and the nymphs, as they grew, had to moult. In the large white butterfly the skin of the caterpillar is soft and can stretch as the caterpillar gets larger. But it cannot go on stretching throughout the life of the caterpillar and you may have noticed that the caterpillar casts its skin every so often.

9   Use a hand lens and focus on the head of one of the caterpillars. Watch the way it feeds.

Q12   How does a caterpillar feed?

Q13   In view of their soft skin, what difference would you expect between the shape of the growth curve of locust nymphs and the caterpillars of the large white butterfly?

10   When the caterpillars are about 40 mm long you should try to look at them several times during the day. The first sign that something is going to happen is that they will stop feeding and seem to get shorter, plumper, and darker in colour. Each caterpillar moves its head from side to side and spins a mat of silk on the leaf and a long thread round the middle of its body attaching it to the silk pad. Watch to see what happens next.

The caterpillar (or larva) does not suddenly turn into a butterfly but something else occurs first. The skin hardens and at the same time the body becomes shorter. The skin then starts to split from the head end and is shed, revealing a chrysalis (or *pupa*) which does not look at all like the larva. The skin of the pupa gradually hardens and you may be able to see the outlines of the future legs, wings, and antennae of the butterfly.

11   Draw one of the pupae and label, on your drawing, the parts of the future butterfly which you can make out.

This rapid change is called metamorphosis. You can read about a similar event in the tadpoles of *Xenopus* and our native frogs and toads in Chapter 5. The change from nymph to adult in the locust was a metamorphosis too.

Q14   What other examples can you think of?

After the caterpillars have formed pupae (pupated) you should place the pupae on a thin layer of damp peat in a plastic box and keep the temperature at about 24 °C. Every day the pupae should be sprinkled with rain water through a fine spray.

After about two or three weeks, the hard skin of the pupa splits and a butterfly with rather crumpled, damp wings emerges. The wings soon straighten out and dry and the butterfly moves them up and down once or twice before taking off on its first flight.

You may be able to watch the butterflies emerging from some of your own pupae. Look at a butterfly and you will see that its body consists of the same parts as the locust's – head, thorax, abdomen, with three pairs of jointed legs and two pairs of wings.

Study *figure 194*, which shows the large white butterfly feeding, and if you have the opportunity, watch a butterfly while it is actually doing this.

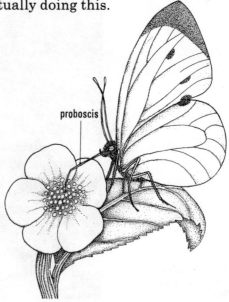

proboscis

Figure 194
The large white butterfly *(Pieris brassicae)*. The proboscis is extended to reach the honeydew in the nectaries of the flower.

Q15　How does the adult butterfly differ in the way it feeds from
a　its caterpillar
b　a locust?

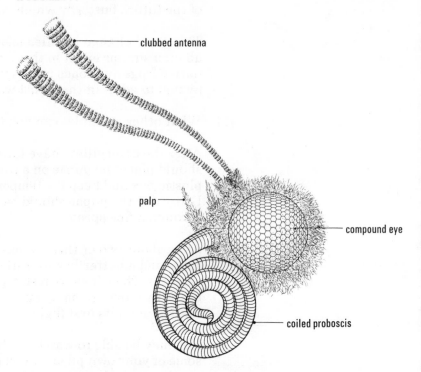

clubbed antenna

palp

compound eye

coiled proboscis

Figure 195
The head of the large white
butterfly, ×15, showing the
proboscis in the resting position.

9.33 The life history of the aphid

Plant lice (or aphids as they are correctly called) are a
common garden pest. There are many types of aphids and
they cause damage by sucking the sap from plants.

Figure 196
The cabbage aphid, *Brevicoryne
brassicae*, on a cabbage leaf.
Photograph, Shell.

Figure 197
Aphids.
a A wingless aphid ×32, using its proboscis to pierce a leaf.
b A 'stem mother' ×32, with young nymphs inside her abdomen.
c Winter eggs of an aphid on a twig, ×2.

Although aphids are very small they do much damage to plants because they occur in very large numbers. One of the commonest is the blackfly or bean aphid *(Aphis fabae)* which is found on broad beans, but if you look in any garden you will find others on plants such as lettuces, cabbages (see *figure 196*), chrysanthemums, and roses (see *plate 1a* on page 229).

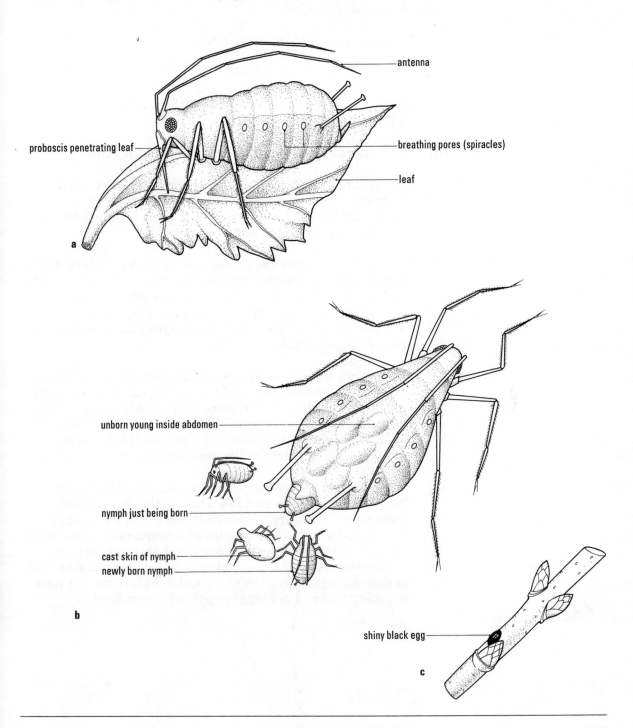

antenna

proboscis penetrating leaf

breathing pores (spiracles)

leaf

a

unborn young inside abdomen

nymph just being born

cast skin of nymph
newly born nymph

b

shiny black egg

c

1 Get an aphid into focus under a hand lens or binocular microscope. Examine it and with the aid of *figure 197* identify the parts of the body.
2 Look at the other aphids.

Q16 Are they all the same size?

Q17 Have any got wings?

3 Look at an aphid from the side.

Q18 Can you see how they feed?

4 Look at one of the large aphids with a dissecting microscope. Focus up and down.

Q19 Can you see anything in the abdomen of the large aphid?

These large aphids are all females and are called 'stem mothers'. Each stem mother can produce 12 or more young aphids in 24 hours and these grow rapidly and can start producing young in 8 to 10 days.

During the summer the females, winged or wingless, go on reproducing generation after generation without a male to fertilize their eggs. Each female may produce a thousand young during the summer but the rate of reproduction slows down in autumn. The ability to produce young without a male is called *parthenogenesis*.

In autumn winged males and winged females are produced and mate. The females fly off to a tree or shrub which does not die during the winter. Here they lay a few eggs on the bark. These eggs have hard shells which protect them during the winter. In spring the eggs hatch into 'stem mothers' which start the life cycle again.

When aphids are feeding they take in a lot of liquid and have to get rid of the surplus. They do this by exuding 'honeydew', a sugary liquid, from the end of the abdomen. This liquid is attractive to ants who frequently stroke the aphids with their antennae to make them produce more 'honeydew'. Aphids are sometimes called 'ant cows' because of this and ants may even transport aphids into their nest in order to have a constant supply of 'honeydew'.

When you clean out the cage containing the stick insects you should look for eggs amongst the droppings.

Figure 198
Eggs of the stick insect, *Carausius morosus* ×7. Note the cap to each egg.
Photograph, Dr Otto Croy.

You will not see stick insects pairing like the locust because males are very rare. The female produces her eggs by parthenogenesis. Collect the eggs and put them in a small dish in the cage. If you keep them long enough (4 to 8 months depending on the temperature and humidity), you will find that some of them hatch into nymphs. The newly hatched nymphs often carry around the egg shell attached to one leg. Do not remove this. It will be shed at the first moult. The nymphs moult several times before they become adult.

Comparing the life histories of insects
In watching locusts you will have seen that at each moult the nymph comes out a size larger. We say that such a type of life history shows an *incomplete metamorphosis* in which there is no abrupt 'changeover' stage.

The life history of the large white butterfly contrasts greatly with that of the locust, for the larva that hatches from the egg is quite different from the adult. The larva changes into a pupa and finally the adult butterfly. This is an example of an insect which undergoes a *complete metamorphosis* in which the pupa is the 'changeover' stage.

Q20 Can you now make further comparisons of the life histories of the insects you have studied by referring to such things as number of stages, how they feed, etc.?

9.4 Insects and man

Insects are sometimes referred to as 'the rival world'. Certainly there is hardly any part of the world in which man is not affected by their activities.

Bubonic plague, which came to England as the Black Death in the fourteenth century and culminated in the Plague of London in 1665, was spread by an insect, the flea *Xenopsylla cheopis*. Plague is actually caused by the bacterium *Pasteurella pestis* which can infect both rats and humans and is transferred from one person to another by the flea. Had the fleas been got rid of the Plague would not have spread.

Figure 199
The flea, *Xenopsylla cheopis*, which carries plague.
Photograph, the Department of Entomology, London School of Hygiene and Tropical Medicine.

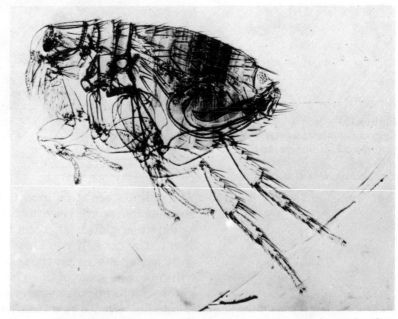

Figure 200
The female yellow fever mosquito, *Aedes aegypti*, 'biting'.
Photograph, Shell.

Great efforts have been made to wipe out mosquitoes in the parts of the world where malaria and yellow fever are common diseases.

Q1 Can you suggest what connection there is between the mosquito and these two diseases?

The photograph in *figure 201* illustrates how, in many hot countries, the locust is a major pest.

Figure 201
A swarm of locusts.
Photograph, Food and Agricultural Organization.

When the eggs hatch, the hoppers march across country eating the vegetation that is in their path. Eventually the adults emerge and form vast swarms which may contain as many as fifty thousand million (5×10^{10}) individuals. These fly considerable distances, and when they settle, consume all the green vegetation in the area. A large swarm may cover a thousand million square metres (10^9 m^2) and consume about a hundred million kilogrammes (10^8 kg) of food per day.

Figure 202 shows how some other insects damage food.

Q2 How many other examples of food damage by insects can
you think of?

However, not all insects are harmful. Many play a useful
role as scavengers, assisting in the processes of decay which
are essential to life. Some control other insects which are
pests, as shown in *plate 1b*.

9.5 Controlling the harmful insects

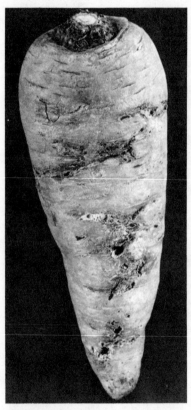

Many of the insects which cause damage and harm man
exist in very large numbers (locusts, mosquitoes, and flies).
Because of this man has spent much time and money on
finding ways of controlling their numbers.

Chemical control

Many substances have been produced which kill insects.
One of the best known is D.D.T. which has been responsible
for getting rid of mosquitoes, and hence malaria, from many
parts of the world. However, it has become apparent that
there are some serious drawbacks to the use of D.D.T. and
similar compounds. It has now been established that
although they do not immediately harm other animals they
do form substances which gradually accumulate in the
bodies of animals. It has also been found that certain strains
of mosquito are now immune to the effects of D.D.T. and
therefore, if they are to be kept under control, new
insecticides or other methods of control are needed.

Biological control

All insect pests are preyed upon by other animals, in many
cases other insects, and this can help to control their
numbers. For instance, aphids are eaten by ladybirds and
their larvae (see *plate 1b*). In some cases, man has been able

Figure 202
A carrot damaged by a carrot fly
(Psila rosae).
Photograph, Shell.

Figure 203
The larva of the flour moth
(Ephestia kuhniella) being
parasitized by an ichneumon fly,
Nemeritis canescens.
Photograph, S. Beaufoy.

Plate 1
a Rose aphids feeding on a rosebud.
Photograph, Heather Angel.
b Adult two-spot ladybirds feeding
on aphids.
Photograph, Shell.

Plate 2
The angle shades moth
Phlogophora meticulosa, well
camouflaged among dead leaves.
Photograph, Heather Angel.

to use this natural relationship to control insect pests.
Greenhouse whitefly is a serious pest of cucumbers which
are grown under glass. The flies collect on the leaves and
weaken and discolour the plants by sucking their sap. We
can control them up to a point by using chemicals but one of
the most effective ways of reducing the numbers of
whiteflies is to introduce a minute tropical wasp. The
female wasps produce their eggs by parthenogenesis and
lay them in the nymphs of the whitefly. When the wasp eggs
hatch, the larvae feed parasitically on the whitefly nymphs
and eventually kill them. A similar sort of relationship is
found between one of the ichneumon flies, *Apanteles
glomeratus*, and the large white butterfly. So far, man has
applied biological control to relatively few pests. However,
scientists are carrying out research which we hope will
mean it can be used more often.

Colour in insects

The colour variations in insects are almost infinite and
many insects are very attractive. In few cases, however, is
the colour purely decorative; it usually protects the insect
from predators in some way.

Opposite page:
Plate 3 *(left)*
The male garden carpet moth,
Xanthorhoë fluctuata.
Photograph, George E. Hyde.

Plate 4 *(right)*
A stick insect (*Carausius morosus*)
on a privet twig.
Photograph, Heather Angel.

A common type of coloration is one which makes the insect
blend with its surroundings. See *plate 2*. Quite often the
general shape of the insect, as well as the colour, matches
the surroundings. You can see this in the case of stick
insects (*plate 4*). Not all of the insect needs to match the
surroundings in order to be camouflaged. It is often enough
if part of it matches the surroundings as long as this breaks
up the outline in such a way that the shape, which stands
out, does not look like the insect. *Plate 3* gives an example.

Introducing living things

Some insects make no attempt at camouflage. On the contrary they have bold colours which make them stand out. This is called warning coloration and serves to tell any would-be predator that the results of an attack would be unpleasant. Common warning colours are reds and yellows combined with black. Wasps (*plate 5*) and bees show warning coloration. In *plate 6*, you can see another example.

Plate 5 *(right)*
Wasps, *Vespa vulgaris*, feeding on jam.
Photograph, Heather Angel.

Plate 6 *(below)*
The cinnabar moth, like wasps and bees, is an example of warning coloration.
a shows the caterpillar and **b** the adult, feeding on ragwort.
Photographs, Heather Angel.

a

b

Introducing living things

In the case of wasps and bees the sting is the obvious threat to the predator but in other cases the colours warn that the insect tastes unpleasant. A number of insects with warning coloration do not have a sting or other means of defence and neither, as far as we know, do they taste unpleasant. These insects simply mimic the warning coloration of other more dangerous animals and obtain protection in this way. A drone fly (*plate 7*) is one such insect.

Some insect colours are apparently designed to confuse predators. The large eye-like shapes that are found on the wings of many butterflies and moths probably do this. See *plate 8.*

Plate 7 *(above)*
A drone fly. It mimics a bee but has no sting.
Photograph, Heather Angel.

Plate 8 *(right)*
The peacock butterfly, *Nymphalis io.*
Photograph, Heather Angel.

If a bird pecks at the peacock butterfly it is likely to attack the brightly coloured 'eye' on the wings. Although this may damage the wing it will be unlikely to harm the butterfly as much as if it pecked at the body. The butterfly can fly with a slightly damaged wing and so may be able to escape. In any case the bird may be put off, as the wing will not form a juicy meal.

Many moths are camouflaged when at rest but as soon as they are disturbed they reveal large, brightly coloured eye spots on the hind wings. This probably confuses the attacker as, at one second, the moth is almost invisible but, at the next, it is very conspicuous. The fact that the colours resemble the eyes of a much larger animal may also help to discourage the predator. (See *plate 9.*)

Some hawk moth caterpillars also have the habit of
revealing two large 'eyes' when they are disturbed.
(*Plate 10.*)

Animal camouflage is a fascinating subject but we must be
careful about the meaning we give to the colours and
shapes of those insects which pretend to be something they
are not. Scientists do not always agree about the exact
explanation of these features but it seems that they must
play an important part in protecting them from enemies.

A strange life cycle: the large blue butterfly

Among the butterflies of Britain the large blue *(Maculinea arion)* occupies an odd position, being one of the most beautiful but one of the least well known. *Plate 11* gives a good idea of its size, and markings, and colour, but to see its real beauty we need the insect itself. Few of us are likely to find the butterfly alive because its homes are now confined to an area of the Cotswolds and a few isolated places in Devon and Cornwall. However, towards the end of last century it occurred in a number of other areas where it is now extinct and, from these, quite large collections were made, many of which can be seen in our museums today. They are well worth a visit. Oddly enough, although the insect is so rare, the sorts of places where it lives vary greatly from glades in woods, to open grassland, and even sea cliffs. However, in all its haunts there occur the wild thyme *(Thymus drucei)* and nests of two common red ants, *Myrmica scabrinoides* and *Myrmica laevonoides*.

The butterflies appear in June and fly only when the sun is shining. One of the reasons why it is difficult to observe them is that during sunny weather they tend to be on the wing only for about three hours, early in the morning. For

Plate 11
The large blue butterfly.
Photograph, S. Beaufoy.

the rest of the day they sit about on the vegetation giving no more than an occasional flutter. To rouse them, the plants must be beaten with a stick. The eggs are laid on the wild thyme, chiefly on the unopened flower buds. Usually, only one egg is laid on each bud and considering the size of the butterfly, these are small, measuring only about 0.5 mm across and 0.3 mm high. They hatch in approximately eight days but the period can vary by as much as three days, depending on the weather. The young larva is a small greenish yellow grub about 0.8 mm long, with an uncanny resemblance to the flower bud on which the egg was laid. It feeds on the flowers of the thyme which provide its only food during the period, lasting about twenty days, of its first three moults. By now the caterpillar is just over 3 mm long. After the third moult, which lasts four days, the large blue larva rests for several hours and then starts wandering about aimlessly. It is fat, grub-like, and clumsy in its movements, and shows no further interest in eating thyme flowers. It seems to be waiting for something. But what?

If a passing red ant happens to come near the larva it immediately takes a great interest in it, caressing it excitedly with its legs and waving its antennae. Then there follows an extraordinary courtship which may last for an hour or more, during which the ant can be seen to suck up little beads of liquid which flow from a honey gland on the back of the caterpillar situated on the tenth segment. The production of honey is stimulated by the activities of the ant which leaves the larva now and then to walk round and round, but each time returns to carry on the 'milking' operation. Meanwhile, other ants may have arrived to share the prize but, apparently, it is a question of first come first served, and they soon disperse, leaving the original captor in possession.

After an hour or more the appearance of the larva begins to change, the segments of the thorax swelling up while the rest of the body remains its normal shape. This seems to be the signal for the next stage in the story, for the ant now seizes the caterpillar in its jaws and carries it off triumphantly to the entrance of its nest, disappearing with its captive below ground. Once in the ants' nest the larva remains there for the rest of its life and another strange thing happens. It ceases to be *herbivorous* (to eat plants) and becomes *carnivorous*, feeding on the young larvae of the ants, which it devours eagerly. It now grows rapidly and as winter approaches it settles down to hibernate among the ant larvae tended by the ants themselves. With the arrival of spring it starts feeding once more and growth is finally completed in early May. Its appearance is now pinkish white and very shiny, its skin giving the impression of being stretched to bursting point. This is probably not far from the truth, for during the time of its captivity among the ants its length has increased fivefold from 3 mm to nearly 15 mm without a further moult!

At this stage the larva changes into a pupa – a whitish object which hangs by a silk pad for a few days from the roof of the nest and then drops to the floor. Here it remains until the perfect butterfly emerges about three weeks later. It crawls along the tunnels of the ants' nest, its wings still folded, and eventually arrives above ground and climbs up a grass stem or onto some other suitable vegetation where its wings can hang down and dry. Within an hour it is ready for its first flight.

So ends the strange story of the life cycle of the large blue butterfly. We may well ask how the partnership of butterfly and ant came about in the first place. The answer must remain no more than a guess. But it is perhaps worth adding that within the group of butterflies called *Lycaenidae* (the 'blue' butterflies) there are a number of other species which will not thrive even in captivity, unless ants are present. How important the ants really are for these species we do not know, but it is unlikely that any other life cycle depends on them quite so completely as that of the large blue.

Figure 204
An ant carrying a caterpillar of the large blue butterfly × 8.
From Frohawk, F. W. (1920) The natural history of British butterflies, *Hutchinson.*

10

Figure 205

The variety of life

You will have come across quite a large number of living things but you may be surprised to know that there are over 950 000 different kinds of animals (just under a million) and about 343 000 different kinds of plants. Clearly, we need some way of dealing with such an enormous variety of living things, besides the plants and animals we know only as fossils.

Many of you will have a stamp collection and even with a much smaller number than a million stamps you will need to sort them into groups or sets. You may decide to do this according to the country which produces them, the subjects of the pictures, or even their shape.

10.1 Putting things into groups

Start by studying a collection of non-living things. Some things to try are assorted shapes like the ones in *figure 205*, a collection of seashells, or some liquorice allsorts!

Q1 Which features seem most useful for this sorting? Don't worry if your ideas are not the same as your neighbour's.

The animals illustrated in *plate 12, a to j*, are all members of the same animal group.

Q2 What is it that they all have in common?

Q3 How many smaller groups can you divide them into?

Q4 Which of their features did you find most useful in helping you to divide them into smaller groups?

You may also name the animals but this is not the most important thing just now. If necessary you will be given help with naming the specimens illustrated in the chapter.

Plate 12

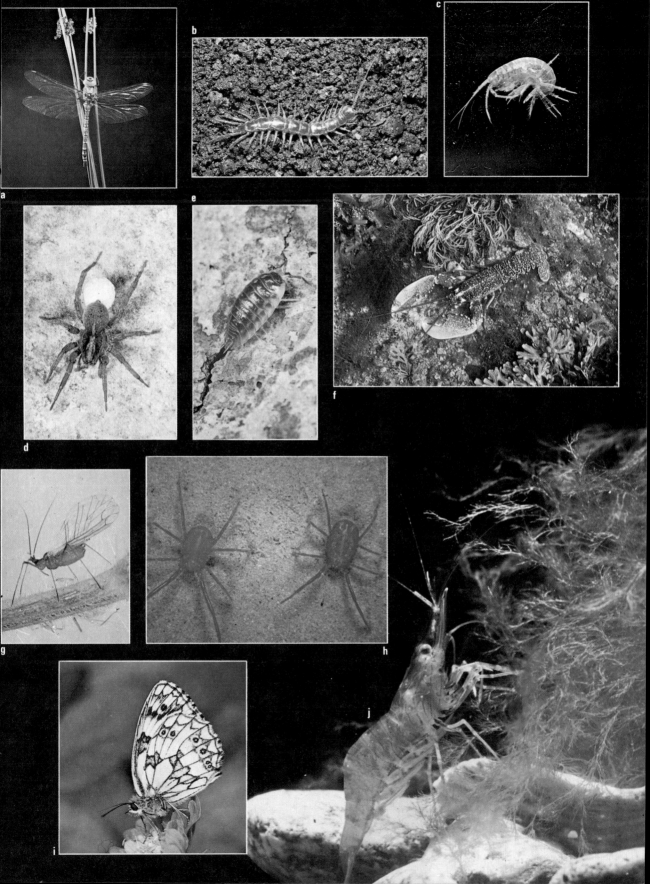

10.2 Groups which biologists use

10.21 Animals with jointed legs

The animals shown in *plate 12* are all members of one group which biologists call the arthropods.

They form one major division of the animal kingdom and can be divided into several smaller groups. The membership of the smaller group depends on features which are different within the larger group.

Table 22 shows one way of dividing up the arthropods. Make a larger copy of the table in your notebook and complete it by adding more features and examples.

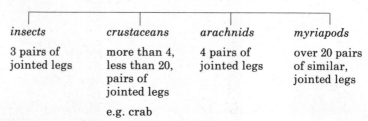

Arthropods
several pairs of jointed legs

insects	*crustaceans*	*arachnids*	*myriapods*
3 pairs of jointed legs	more than 4, less than 20, pairs of jointed legs	4 pairs of jointed legs	over 20 pairs of similar, jointed legs
	e.g. crab		

Table 22

The groups which biologists have devised over the years may be based on features of plants and animals which we can see with the naked eye, a lens, or a microscope, but many of the clues come from a study of development and internal structure. Also, during the millions of years in which many groups have existed, changes have taken place in the basic patterns of their bodies. Some plants and animals are quite easy to sort correctly; others are rather more difficult. The photographs in *plate 13, a to f*, are all of arthropods. Each has at least one clue visible, sometimes more.

Q1 In which group, within the arthropods, would you place each example?

Q2 Which clues did you consider important in each case?

Q3 Have you had to revise your list of group features?

Plate 13

Introducing living things

Figure 206
A 'pie-chart' showing the proportion of arthropods to all other animals.

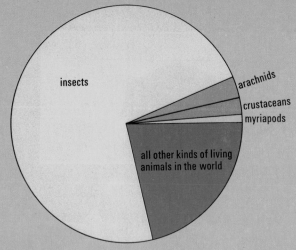

insects

arachnids

crustaceans

myriapods

all other kinds of living animals in the world

Plate 14

a

b

c

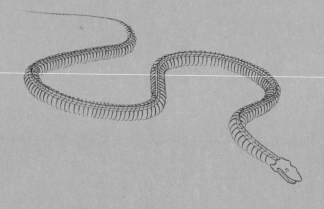

You may have wondered why we have begun our study of the variety of life with one particular group from the animal kingdom. Study the 'pie-chart' in *figure 206* and suggest why it may have been done. Now study the photographs (and drawings) in *plate 14, a to f*.

Q4 Why has a skeleton from inside the body been shown in some cases and not in others?

Q5 One other animal shown also has an internal skeleton. Which one is it?

Q6 Study the shape of the animal shown in *plate 14b*. Why is it 'odd man out'?

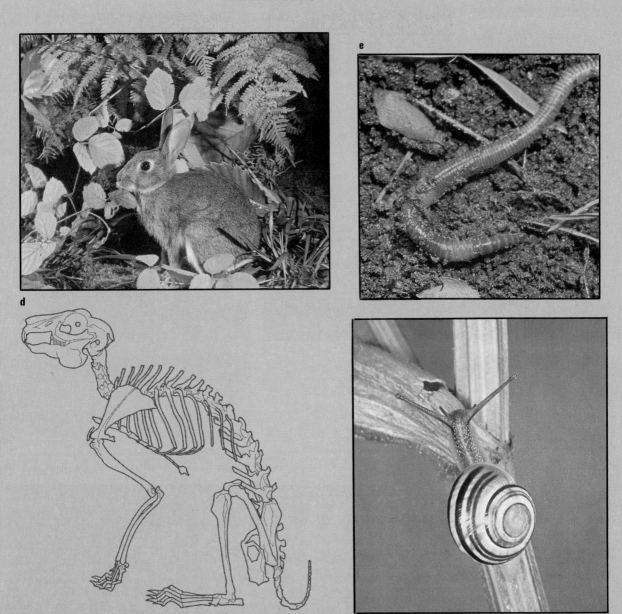

10.22 Animals without a backbone

Study *plate 15*, *a* to *k*, and any other photographs and specimens provided. All are from groups which do not have a backbone. That is, all are *invertebrates*.

Q7 How many arthropods can you find?

Q8 How many other groups can you divide them into?

Q9 Which features are most useful for helping you with this sorting?

Plate 15

a

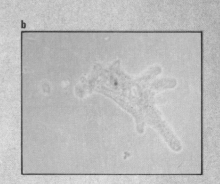
b

c

d

e

f

g

i(1)

i(2)

h

j

k

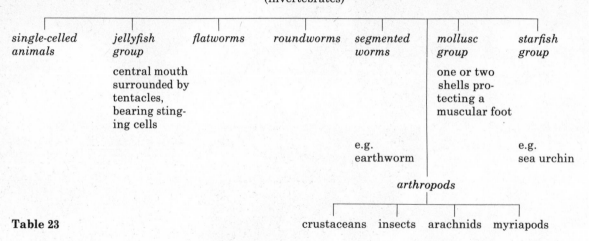

Animals without backbone
(invertebrates)

single-celled animals	jellyfish group	flatworms	roundworms	segmented worms	mollusc group	starfish group
	central mouth surrounded by tentacles, bearing stinging cells				one or two shells protecting a muscular foot	
				e.g. earthworm		e.g. sea urchin

arthropods

crustaceans   insects   arachnids   myriapods

Table 23

Table 23 shows the way in which biologists may divide up the animals without a backbone. Make a larger copy of the table in your notebook and give more features and examples.

Q10   *Table 23* gives some names which are the English versions of the scientific names, e.g. 'arthropods'. In other cases, e.g. 'segmented worms', this has not been done. Can you suggest why?

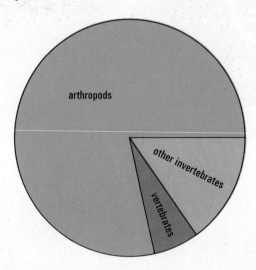

Figure 207
Another version of the 'pie-chart' shown in *figure 206*.

The photographs and drawings in *plate 16, a* to *h*, show more examples of invertebrates. Try to place each one in its correct group. All have at least one clue to help you.

Plate 16

a

b(1)

b(2)

c

d

e

f

g

h

10.23 Animals with a backbone

So far we have only studied the animal groups whose members have an outside skeleton or none at all. Many animals, like the rabbit, slow worm, and frog shown in *plate 14, d, c,* and *a,* have an internal skeleton, mostly of bone. This skeleton includes a vertebral column and another name for the group is therefore *vertebrates.*

Sort out the animals shown in *plate 17, a* to *j,* with any other specimens or photographs provided, into smaller groups, as before. Give your reasons for the way you group them.

Table 24 shows the main groups within the vertebrates Make a larger copy of this table in your notebook and complete it by adding more features and examples.

Plate 17

a

b

c

d

e

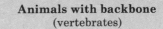

Animals with backbone
(vertebrates)

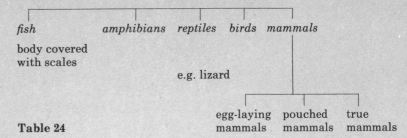

fish   *amphibians*   *reptiles*   *birds*   *mammals*

body covered
with scales

e.g. lizard

egg-laying   pouched   true
mammals   mammals   mammals

Table 24

f

g

h

i

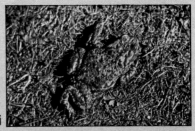

j

Plate 18

Now place the animals shown in *plate 18*, *a* to *g*, in the correct group. Once more you will have to look closely at the photographs and drawings for clues.

a

b

c

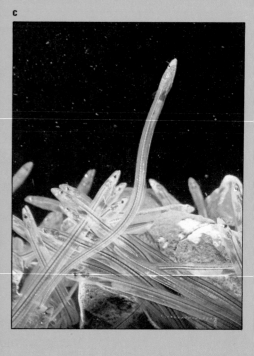

d

f

e

g

10.24 Plants

So far we have studied only members of the animal kingdom, but plants may be divided into groups in the same way, although the kinds of features which help sorting may be rather different. *Plate 19*, *a* to *k*, shows a wide range of plants. Once more, use your own ideas to sort them into smaller groups, giving your reasons.

Plate 19

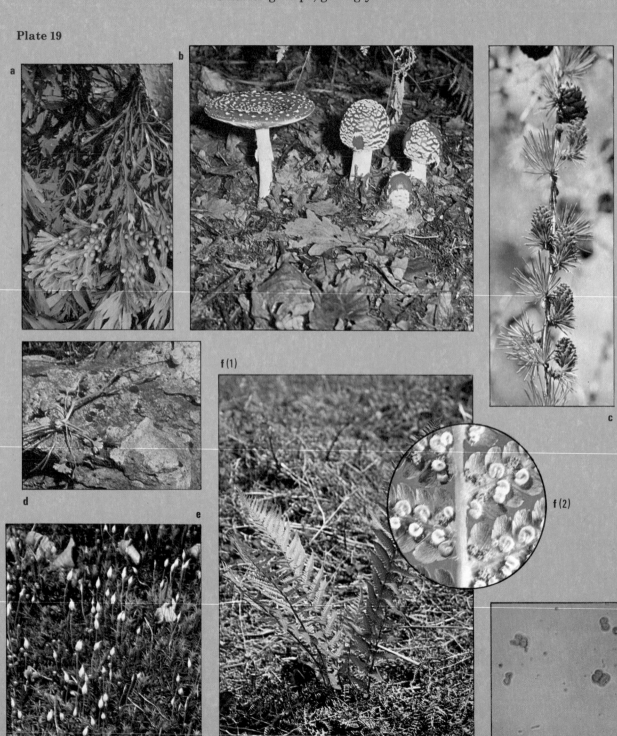

(1)   h(2)   i

j   k

Table 25 shows the main plant groups. Make a larger copy
of this table in your notebook and try to complete it by
adding more features and examples.

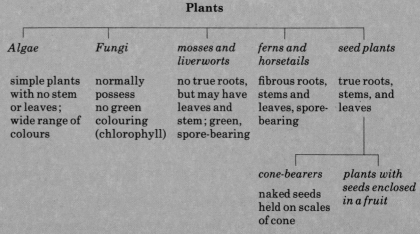

Plants

Algae	Fungi	mosses and liverworts	ferns and horsetails	seed plants
simple plants with no stem or leaves; wide range of colours	normally possess no green colouring (chlorophyll)	no true roots, but may have leaves and stem; green, spore-bearing	fibrous roots, stems and leaves, spore-bearing	true roots, stems, and leaves

cone-bearers	plants with seeds enclosed in a fruit
naked seeds held on scales of cone	

Table 25   N.B. *Lichens* are made up of thread-like Fungi and one-celled green Algae.

Plate 20

Plate 20, a to f, shows other plants. Study them carefully and decide to which plant group they belong.

a(1)

a(2)

b

d

f

e

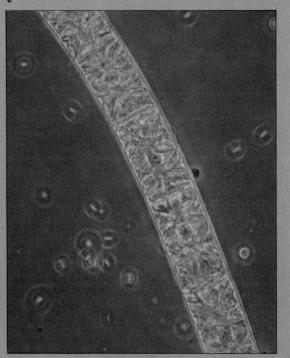

10.25 Fossils

The plants and animals you have studied so far have all been those living now but we know many plants and animals only as fossil remains. These can be of various kinds. Usually the tissues have become impregnated with hard minerals and we say they have become petrified. Sometimes we see the petrified remains, sometimes only an impression. *Figure 208* contains pictures of fossils or casts, or artists' reconstructions of organisms that are no longer living. Into which plant or animal groups would you put them?

Figure 208

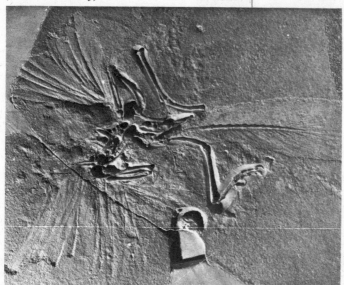

a(1) A cast

a(2) An artist's reconstruction

b

c

d

e

Figure 208 *(continued)*

f

10.26 Plants or animals?

Almost all living things can be grouped quite easily in either the plant or animal kingdom.

Make a list of any ways in which plants and animals are similar.

Now make a table which shows any differences that you have noticed.

Q11 If you have studied Chapter 7, try to decide where you would classify bacteria. Would you put them with plants or with animals? Or neither?

Q12 There are many more animal fossil remains than there are plant fossils. Can you suggest why?

10.27 Larger and smaller groups

Within the animal kingdom there are several large groups which are called phyla. The arthropods form one phylum.

Q13 Can you name seven other phyla after studying *table 23* (page 246)?

Each phylum is sub-divided into smaller groups called classes, e.g. insects or crustaceans. There may be further smaller groups until the two special scientific names of a species are reached. An example is given in *table 26*. Some Latin names have been given on this occasion.

Kingdom	animal
Phylum	arthropods
Class	insect
Order	Lepidoptera (scaly winged)
Genus	*Pieris*
Species	*brassicae*
Common name	large white butterfly

Table 26

Figure 209
Four common wracks.

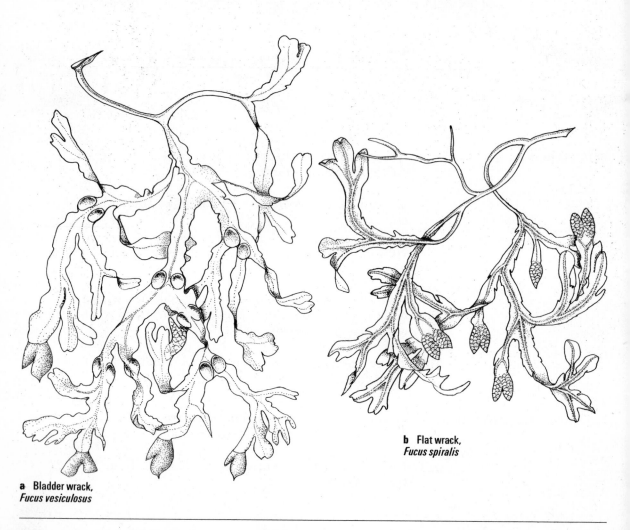

a Bladder wrack,
Fucus vesiculosus

b Flat wrack,
Fucus spiralis

Introducing living things

The 'surname' *Pieris* is shared by several similar butterflies. Thus, *Pieris rapae* is the small white butterfly and *Pieris napi* is the green-veined white.

Note that the generic name or 'surname' is given a capital letter; the specific name that follows is written without one; they are printed in italic and when written they are underlined.

We are members of the genus *Homo* and the species *sapiens*. Fossil remains of other species of the genus *Homo* have also been found including *Homo erectus* in Java and China. (*Homo erectus* is also called Java man and Pekin man.)

Plants are divided up into smaller groups, in a similar way.

The Algae, four of which are shown in *figure 209*, make up one phylum of the plant kingdom. We have already met several different coloured Algae and this variation in

Figure 209 *(continued)*

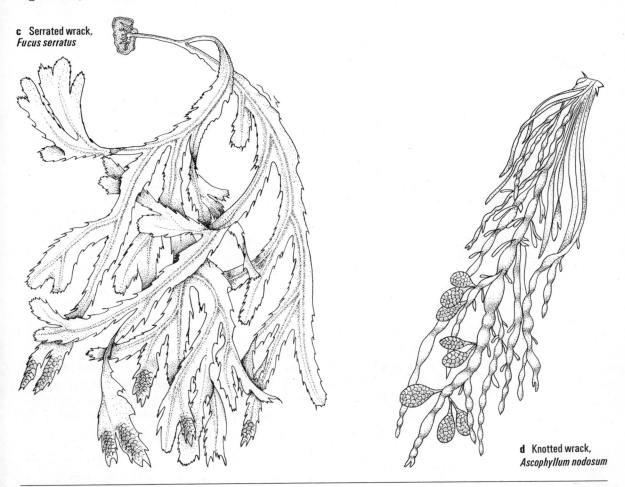

c Serrated wrack, *Fucus serratus*

d Knotted wrack, *Ascophyllum nodosum*

colour is used to divide up the phylum into several classes including red Algae, brown Algae, green Algae. Within the brown Algae we have met the bladder-wrack whose scientific name is *Fucus vesiculosus*.

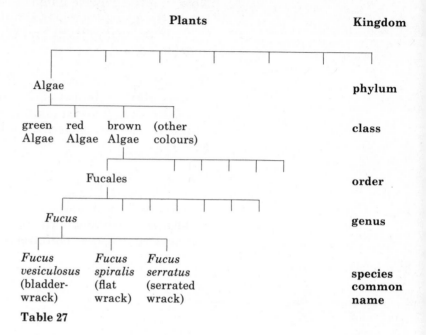

Table 27

Many amateur biologists prefer to use 'common' or English names wherever possible and many of these are delightful. *Arum maculatum* is known in different parts of the country as 'lords and ladies', 'cuckoo pint', 'wake-robin', and 'Jack in the pulpit'.

In England *Endymion non-scriptus* is the bluebell. But the 'bluebells of Scotland' are our harebells, *Campanula rotundifolia*. We have used bladder-wrack as the common name for *Fucus vesiculosus* but in some parts of the country the common name might refer to another brown seaweed, *Ascophyllum nodosum (figure 209)*.

To be precise, scientific names are essential, especially when discussing plants and animals from different areas or countries. The name *Fucus vesiculosus* refers to one particular plant whether in England or Japan!

The Swedish naturalist Linnaeus was the first person to try to classify all living things systematically, by giving each a name for its genus and a name for its species. There is more about Linnaeus in the Background reading at the end of the chapter.

Figure 210 *(left)*
Arum maculatum, wild arum.

Figure 211 *(centre)*
Endymion non-scriptus, the
bluebell.

Figure 212 *(right)*
Campanula rotundifolia, the
harebell.

10.3 Naming things

By now you have discovered that there are so many living
things on the Earth, and there were so many more which
have become extinct, that it would be quite impossible to
find out about all of them, let alone find out all their names.
But it often happens that, when carrying out an
investigation, you want to know the name of a particular
animal or plant. You may want to make a list of all the
animals and plants you could find in a garden, or you may
want to distinguish between one animal and another one
which looks very like it. If you have studied earthworms you
may have come across a number of different kinds, each
having two Latin names like the common earthworm,

Lumbricus terrestris (*Lumbricus* = its genus and *terrestris* = its species). It is fairly easy to distinguish between two members of the cat family – a lion and a tiger, for instance. But if you had to decide whether an earthworm was the common earthworm or the rather similar long worm (*Allolobophora longa*) you would have to make a very careful examination in order to find out which one it was. In other words, to distinguish between animals or between plants which look rather alike, you may have to look at their characteristics much more closely.

To find out the name of a plant or animal you could use a book with good illustrations and just turn through the pages of pictures until you came to one which seemed to be right. This is often a good way of identifying something, provided the pictures are detailed and accurate, and provided you make a careful comparison between the specimen and the picture. But, even so, there are often cases where the picture will not give you the answer, because it may not show the important features clearly. In any case, it would be expensive to buy enough books with good illustrations to name all the specimens you might find.

Because of these difficulties, most books which enable us to find out the name of things make use of what are called 'keys'. Such a key consists of a series of questions which, when you have answered them, lead you to the name of the specimen. This sounds quite easy, and you will find that it is, once you have had some practice.

10.31 Making a key

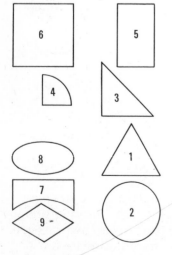

Figure 213

At the beginning of the chapter, you looked at a collection of non-living things to introduce the idea of sorting. The same collection can be used to practise making a key. Look for simple, clear-cut features which divide the collection into smaller and smaller groups, until each object is separated from the others.

Use the key you have made to identify shape number 9 in *figure 213*.

Q1 What does this tell you about keys of this kind?

There are many other collections for which you could make a key. You could, for example, make a key by which a new member of the staff would be able to identify each person in the class. If you do this, try answering questions 2 to 4.

Q2 When you made the key, which features did you find most useful and reliable?

Q3 What groups in the school could this key be used for?

Q4 Would your key be as useful next term?

10.32 Using keys

The keys you have made so far have been the so-called 'spider' keys. These are very useful for beginners but for a large number of specimens they take up a lot of room. Keys can be arranged rather more compactly. *Table 28* is a short example of a numbered key.

Table 28
A simple key to some common trees, using the features of their winter buds.

1	Buds more than 4 times as long as broad	2
	Buds less than 4 times as long as broad	3
2	Buds where cigar-shaped leaf scar nearly surrounds base of bud	beech
	Buds not cigar-shaped. Leaf scar much smaller	birch
3	Buds smooth, green-brown or black	4
	Bud scales red-brown, sticky. Leaf scar horse-shoe shaped	horse-chestnut
4	More than 2 green-brown bud scales	lilac
	Only 2 black bud scales	ash

Q5 Using the key in *table 28* identify the twigs illustrated in *plate 21*.

Plate 21

a b c d e

Q6 Can this key be used to identify the twig illustrated in *plate 22*?

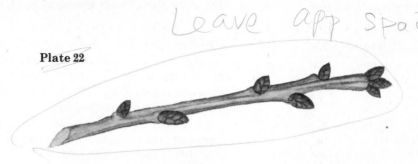

Plate 22

A key can only be used to identify a specimen if that specimen was included in the collection used to make the key. When an answer is reached using a key, further descriptions or illustrations should be studied to make sure that the answer is the correct one.

This is particularly important if the specimen is rare or foreign. Many keys, e.g. trees, include only common, native specimens. It may not be possible to identify more unusual trees met in gardens or parks, as they would not have been included when making a key of common, native trees.

Keys which identify a selected collection of specimens in this way, using easily observed features, are called 'artificial keys' because they are not related to the natural classification patterns studied earlier in the chapter.

Background reading

How Carolus Linnaeus wrote 'the great alphabet of nature'

In this chapter you have been reading about the way in which living things are described and classified. You will have seen that a *system* is used to divide animals from plants and to separate one type of plant or animal from another. *Table 26* shows a good example of the way the system works.

Ever since the days of the Ancient Greeks men have sought to impose order on the chaos of nature. Today, the ways in which they tried to do this seem very confusing to us largely because we believe many of the notions they held to be wrong. The orderly system we use today, however, had its origin almost 250 years ago, in the work of one man, Carl von Linné (known to us as Linnaeus from the Latin version of his name, which his family assumed). Linnaeus, who was born in 1707, was the son of a Lutheran minister in Sweden, and seemed destined to follow his father into the ministry. His interests, though, lay in another direction, and there were family squabbles particularly with his mother who, a clergyman's daughter herself, seemed to think her son would be breaking a sacred trust if he did not go into the ministry. Eventually he persuaded his parents to let him enrol at a nearby university to study medicine. By the time he was 22 he had already published an essay which earned him praise from his teachers and led to an invitation to lecture to audiences of fellow students. In this essay, written in the flowery language so popular at the time, Linnaeus described how plants were organized on the basis of their reproductive parts, and how the way in which the stamens and carpels in flowers were built made reproduction possible. Linnaeus called his essay 'The marriage of the flowers' and it was the real beginning of his famous system of classification, based largely on the sexual parts of plants, which dominated botany for more than a hundred years.

Linnaeus had by now become fascinated by the idea that living things could be named and classified in a concise and understandable way. As he wrote later '... the first step of wisdom is to know these bodies, and to be able, by those marks imprinted on them by nature, to distinguish them from each other, and to affix to every object its proper name.'

Figure 214
Linnaeus in Lapland dress.
Photograph, Ronan Picture Library.

'These are the elements of all science, this is the great alphabet of nature: for if the name be lost, the knowledge of the object is lost also; and without these, the student will seek in vain for the means to investigate the hidden treasures of nature.'

For Linnaeus the next step in the search was a journey of discovery into Lapland, the northern part of Sweden, which was virtually unexplored and very primitive territory. How primitive can be seen in this description of the people by the German poet Heinrich Heine: 'In Lapland the people are dirty, flat-headed, wide-mouthed and small; they huddle round the fire, frying themselves fish, croaking and shrieking.' After an epic journey of 4000 miles he returned to the university late in 1732, with a journal full of descriptions of the country and its inhabitants, and with an enormous collection of plant specimens. The trip had been well worth while for the young Linnaeus (he was still only 25) and he kept vivid memories of the adventures and hardships he had experienced for the rest of his life. On later occasions, lecturing to students or discussing his system with members of learned societies, he needed little persuasion to dress up in the peasant costume he had worn on the journey. He even posed for his portrait wearing it, as you can see in *figure 214*.

By this time Linnaeus was celebrated and famous throughout Europe. The eighteenth century was an age in which the discovery of new lands went hand in hand with an intense interest in nature and all its products. The rich and educated in many countries had developed a passion for collecting plants and many new species had been discovered. Up to the beginning of the seventeenth century only 6000 varieties were known. In the next hundred years almost 12 000 more were added by these enthusiastic amateur botanists, including species from the four corners of the Earth. In one genus Linnaeus was able to describe species that had been collected not just from Europe but from Crete, Siberia, and even Tierra del Fuego on the southernmost tip of South America. So many rival systems of classification were then in use that it was rapidly becoming impossible for doctors, who needed to be able to recognize and find certain plants for their medicines, and for collectors, to find their way through the maze of conflicting names and groupings. Because there was so little real system about ways of naming and classifying, Linnaeus' simple division, into genus and species based on common characteristics, was soon seen to be a great breakthrough in the science of botany. An example of the conciseness and simplicity of his work is his way of naming the red maple tree. While Linnaeus was able, within the

Introducing living things

context of his system, to refer to it merely as *Acer rubrum* (red maple), his predecessors' only way of making sure there was no confusion was to give it a name which included *all* its particular characteristics, so *Acer rubrum* becomes *Acer Americanum, folio majore, suptus argenteo, supre viridi splendente, oribus multis coccineus* (American maple, with big leaves, silvery underneath and a lustrous green above, and also having many touches of scarlet). Just think how difficult it must have been for collectors to label their specimens and produce orderly catalogues!

Figure 215
A botanical foray, led by Linnaeus.
Photograph, © *George Rainbird, Ltd, 1971, by courtesy of the Linnean Society.*

Linnaeus continued his travels soon after his return from Lapland. First he went to Holland, the intellectual centre of Europe at the time, where he was able to have several manuscripts published. These included the first edition of *Systema Naturae* which was only 14 pages long. By the 12th edition in 1766 it had grown to three volumes and 2300 pages! Then, hearing of the journey an Englishman, Sir Hans Sloane, had made to the West Indies he determined to see Sloane's garden in Chelsea. He might have stayed in England for he found much to interest him during his visit. The great botanist of the previous generation, John Ray, was English, and Englishmen were enthusiastic collectors and keen to hear about the new science of systematic botany.

Linnaeus was still only 30 years old yet he was already being hailed as 'the prince of botanists'. Like Lavoisier, the great French chemist who lived at about the same time, Linnaeus, with a stroke of his pen, had swept away all the confusing and misleading names which had stood in the way of the development of his subject. The outlines of

Linnaeus' system are still in use today although the basis of common sexual characteristics on which it was founded has, since Darwin's evolutionary theory, been heavily modified. At the time the sexual descriptions were greeted with horror by many scientists. One Russian even denounced the system as 'lewd' and asked: 'Who would have thought that bluebells, lilies, and onions could be up to such immorality?' Linnaeus got his own back for this criticism in a characteristic way. His critic's name was Siegesbeck and Linnaeus, in revenge, named a small, unpleasant weed, *Sigesbeckia*, in his honour.

Linnaeus perfectly represented his own belief that 'It is the exclusive property of man, to contemplate and to reason on the great book of nature. She gradually unfolds herself to him who, with patience and perseverance, will search into her mysteries.'

After he died in 1778, his wife sold his collection of papers and specimens to a young Englishman, James Edward Smith. She had first offered them to Sir Joseph Banks, the great English naturalist and founder of Kew Gardens, but, surprisingly, he thought his own collection too big to add Linnaeus' treasures to it, and suggested that Smith should make an offer, which he did, successfully. When they realized what a prize had been lost to them, Swedish academics were shocked and angry. In fact, if the King of Sweden had not been away at the time he would very probably have stepped in and prevented the removal of the collection. However, rage and protest were to no avail and the collection now belongs to the Linnean Society in London where Linnaeus' neat handwriting and childlike drawings and sketches can still be seen.

Ad prg. 96 Fig. I.

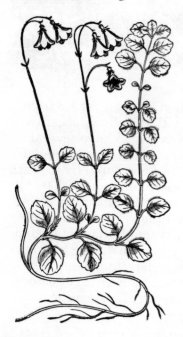

Figure 216
A woodcut of *Linnaea borealis*, named after Linnaeus.
© *George Rainbird, Ltd, 1971.*

The colour photographs in this chapter were provided as follows: Heather Angel: *plates 12a, b, f,* and *j, 13a, b, c, d,* and *e, 14a, b, c,* and *e, 15a, c, g, h,* and *i (1* and *2), 16a, d, f,* and *h, 17c, j, e,* and *f, 18b* and *c, 19b, c, e, f(1)* and *(2), h(1)* and *(2), j,* and *k,* and *20a (1* and *2), b,* and *c*; Australian News and Information Bureau: *plates 17h* and *18a*; S. C. Bisserôt/Bruce Coleman Ltd: *plates 12c* and *i* and *18d*; Jane Burton/Bruce Coleman Ltd: *plates 14d, 15j,* and *17b*; N. A. Callow/Natural History Photographic Agency: *plate 13f*; John Clegg/Tourist Photo Library: *plate 12c*; R. J. Corbin: *plate 20f*; Department of Medical Protozoology, London School of Hygiene and Tropical Medicine: *plate 15f*; M. T. Feesey: *plate 20d*; D. W. Fry, Commonwealth Mycological Institute: *plate 19i*; Harris Biological Supplies Ltd: *plates 19g* and *20e*; Peter Jackson/Bruce Coleman Ltd: *plate 17a*; Russ Kinne/Bruce Coleman Ltd: *plates 16e* and *18f* and *g*; R. K. Murton/Bruce Coleman Ltd: *plate 17i*; Norman Myers/Bruce Coleman Ltd: *plate 17g*; The Pharmaceutical Society of Great Britain, from North, 1967, *Poisonous plants and fungi in colour*, Blandford: *plate 19d*; M. W. F. Tweedie: *plate 12d, g,* and *h* and *plate 14f*; M. W. F. Tweedie/Natural History Photographic Agency: *plate 16c*; M. I. Walker: *plate 15b*; Derek Washington/Bruce Coleman Ltd: *plate 17d*; Dr Douglas P. Wilson: *plates 15d* and *e, 16b* and *g, 18e,* and *19a.*

Introducing living things

Man and his environment

By now, you will have investigated a number of different animals and plants and found out how they reproduce and how their young develop and grow up. You will realize that in order to grow up successfully, a plant or animal must live in the right sort of place. An earthworm, for instance, is the right shape to live in the soil. Its body is constructed in such a way that it can move easily through the soil and feed on the plentiful supply of decaying vegetable matter, or humus, which is there. The soil, therefore, is the earthworm's natural environment or what is called its *habitat*.

All animals and plants have their particular habitats where they can live successfully and multiply, and you may have investigated some of these if you studied Chapter 1.

Some voles, for instance, live in holes in a bank. There they make their nests and rear their young close to their supply of food, that is, insect larvae and wild fruits. Similarly, individual plants live only in certain habitats. Ferns and mosses are found in damp places, while species such as ivy climb up a wall or a tree trunk as they grow.

We must not forget that living things are themselves part of a habitat. One animal or plant may be dependent upon another for its food or for assistance in its reproduction. This is the case in many flowers which depend upon insects to transfer pollen from one flower to fertilize another.

The great naturalist Charles Darwin discovered about one hundred years ago that red clover could only be pollinated by bumble bees. Because they have longer tongues than some insects, bumble bees can reach the nectar at the bottom of the flowers upon which they feed. In a field of clover, bank voles often destroy the grass nests and eat the larvae of the bumble bee, and then use the nests to rear their own young in. The voles are, in turn, eaten by birds

Figure 217
Links in a food chain.

of prey such as the owl, which fly over the field. Thus, in a single field habitat, there is a *community* of plants and animals each depending upon the other for its supply of food. We can summarize the situation as:

red clover → bumble bees → bank voles → owls

Each separate source of food is like a link in a chain and the process is known as a *food chain*. Another example of a food chain is:

cabbage → large white butterfly → thrush → kestrel

Q1 Can you work out some more food chains based on your own observations? Watching animals in your own garden, the school grounds, or a local park, field, or wood will help you to do this.

Q2 What kind of living organism is to be found at
a the beginning of a food chain
b the end of a food chain?

By this time you may have realized that simple, straight food chains rarely exist. For instance, thrushes do not eat large white butterfly caterpillars only. They eat a variety of creatures including snails. Neither are cabbages eaten only by large white butterfly caterpillars; greenfly, slugs, and pigeons also feed on them.

A more accurate summary will look like this:

```
pigeon        greenfly
  ↑         ↗
cabbage → large white butterfly caterpillar → thrush → kestrel
          ↘
            slug
```

This is still not quite the complete picture but perhaps you can see that rather than use the term 'food chain' it is often better to talk about a *food web*.

Q3 What other 'strands' might be added to this example of a food web to make it more nearly complete?

Q4 What other food webs based on, for example, the organisms associated with a lawn, a small tree, or a pond can you work out?

Figure 218 gives another example of a food web. Study this web or any other one you have worked out and answer the following questions.

Q5 How many simple food chains are there within the web? Write out each one separately and notice how they connect with each other.

Figure 218
A food web.

Q6   What do you notice about the number of links in any one food chain?

Q7   In the example given in *figure 218*, how would the number of small birds compare with the number of hawks?

Q8   How do you think the numbers of organisms at the different stages of a food chain compare with each other?

Q9   How do the sizes of the animals at the different stages of a food chain compare with each other?

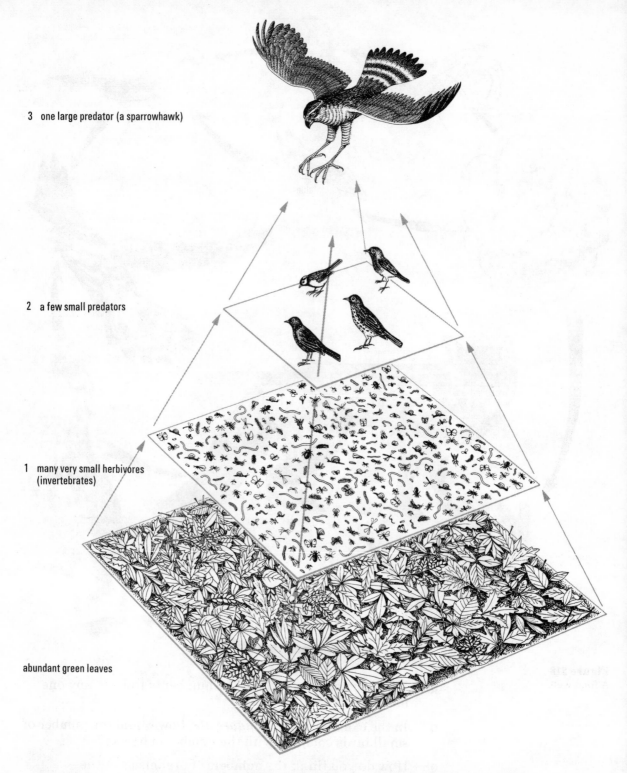

3 one large predator (a sparrowhawk)

2 a few small predators

1 many very small herbivores
(invertebrates)

abundant green leaves

Figure 219
A pyramid of numbers.

Figure 219 shows another way of presenting the information given by a food web. You can see why this is called a *pyramid of numbers*.

Introducing living things

Q10 How do the organisms found at the base and top of the pyramid compare with those at the beginning and end of a food chain?

Chapter 7 discusses how microbes are useful in causing the decay of dead plant and animal matter. In Chapter 9, insects are mentioned for the useful role some of them have as scavengers, so helping in the process of decay.

Q11 How and where do scavengers and microbes fit into a food web and a pyramid of numbers?

Although man has not been shown in the examples given of food chains or in the pyramid of numbers, you may have worked out examples which include man. If you have not already done so, work out an example of a food chain in which man does appear.

Q12 What effect would it have on man if one of the layers in his pyramid of numbers were destroyed?

From this it should be obvious that it is only possible for the population of any habitat to survive for generation after generation if the numbers of plants or animals at each link of the food chain do not increase or decrease too much.

Q13 Why is it so important to disturb a habitat as little as possible and to return animals to their habitat when carrying out investigations such as those suggested in Chapter 1?

When the numbers of animals and plants in a particular habitat remain approximately the same over a period of time we have what is called a *balanced community*. But what happens if the balance should be upset by altering one or more of the links in the chain?

11.2 How man affects the balance

To return to the first example of a food chain, suppose a gamekeeper kills off all the owls and other birds of prey in the area. The enemies of the bank vole will then be reduced and the voles can multiply unhindered. With so many to feed, their food will become scarce and the voles will attack and destroy more nests of bumble bees, eating their larvae. This, in turn, will mean that fewer clover flowers will be pollinated and fewer seeds produced. Soon there will be no clover in the field, and the community will cease to exist.

Thus, by destroying owls, man can upset the balance in a field habitat. But he can do this in other ways as well.

The photographs in *figures 220, 221, 223* (page 277), *225* and *226* (page 278), *228* and *229* (page 280) show examples of how man has, in various ways, affected the *balance in nature*.

Q1 Look at *figure 220*. How has this type of pollution of a river happened?

Q2 What effect may it have on the balance of the plants and animals living in the river?

Q3 How would you set up an experiment to try to find out what effect detergents have on plant and animal life?

Figure 220
Near Gunthorpe Weir in Nottinghamshire; detergent foam on the river Trent. (Since this photograph was taken, much has been done to free the river from pollution of this kind.)
Photograph, Nottingham Evening Post and News.

274   Introducing living things

Figure 221
a The river Itchen, in Hampshire.
Photograph, John Tarlton.

b The river Stour, in Kent.
Photograph, Kentish Gazette.

One of the most attractive features of any chalk stream is the variety and abundance of the plants living in it. The animal life is also plentiful and will include insect larvae such as mayfly nymphs, crustaceans such as the small freshwater shrimp *Gammarus pulex* and the much larger crayfish, as well as a variety of fish. It is not surprising that these rivers provide some of the best fly fishing in the country.

Both of the rivers shown in *figure 221* are examples of the chalk stream habitat.

Q4 What differences can be seen in the appearance of these two rivers?

Two possible causes of river pollution are sewage effluent and drainage water, containing fertilizers, from farm land. These substances will act as fertilizers in water just as they do on the land and one of the results can be a very rich growth of freshwater Algae of the type called flannelweed.

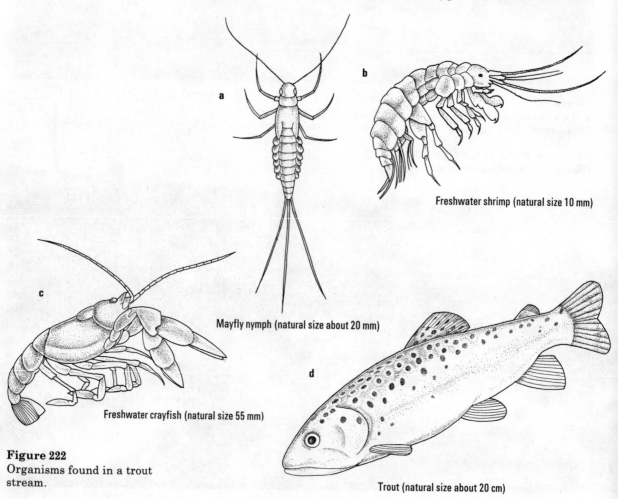

a

b

Freshwater shrimp (natural size 10 mm)

Mayfly nymph (natural size about 20 mm)

c

Freshwater crayfish (natural size 55 mm)

d

Trout (natural size about 20 cm)

Figure 222
Organisms found in a trout stream.

Introducing living things

Flannelweed is dark and slimy and in large quantities it has an unpleasant smell. It grows over and among the lovely stream plants, so that these cannot get enough sunlight and die. As you already know, micro-organisms will help in the decay of the dead plants but they use up oxygen to do so. As a result, the animals begin to die as well. So, eventually, the typical stream life will disappear.

Q5 What do you suggest might be the cause of the differences in the appearance of the river Itchen and that of the river Stour?

Q6 What steps would you suggest should be taken to prevent such pollution of a river?

Figure 223 *(right)*
Shrub growth on a part of Box Hill, Surrey, in 1959. Previously, the land had been typical of chalk grassland.
Photograph, The Nature Conservancy.

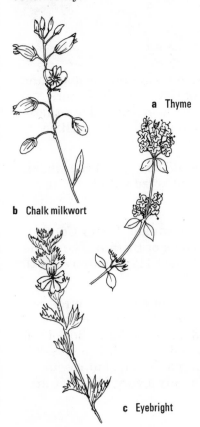

a Thyme

b Chalk milkwort

c Eyebright

Figure 224
Typical plants that grow on chalk grassland.

Box Hill in Surrey is an example of chalk grassland, known for its short grasses and many small flowers such as thyme, chalk milkwort, and the eyebrights.

The photograph in *figure 223* shows what happened to the grassland in the 1950s after myxomatosis had killed most of the rabbit population.

Q7 How was the balanced community upset in this case?

Figure 225 *(right)*
Photograph, Daily Telegraph.

Figure 226
A badly oiled guillemot *(Uria salge).*
Photograph, Eric Hosking.

Figure 225 shows the notice which stands on the mainland of Scotland opposite the island of Gruinard. It is there because germ warfare experiments were carried out there during the Second World War and anthrax may still be harboured in the soil.

Q8 Is the fact that human beings should be kept off the island, for at least one hundred years, connected with keeping a balance in nature? How?

In March 1967 the 118 000-tonne oil tanker *Torrey Canyon* was wrecked off the south-west coast of Britain. The crude oil which was spilled reached at least 112 kilometres of the British coast.

The disastrous effect of oil on the plumage of sea birds can be seen in *figure 226.* Oiled sea birds cannot feed or keep warm and they cannot swim or fly, and so thousands of sea birds died after the *Torrey Canyon* disaster. It is impossible for oiled birds to survive without man's help, but rescue operations are difficult and often only a very few birds are saved and returned to the sea.

Q9 In what other ways might crude oil pollution of the seas upset a balanced community? Remember that chemicals, such as detergents, may have to be used to remove the oil.

Figure 227
The peregrine falcon *(Falco peregrinus).*
Photograph, Eric Hosking.

The peregrine falcon is a particularly fine bird, especially if seen in flight. It has never been a very common bird but has bred around our coasts for many years. Recently, a count showed that in the last ten years the number of breeding pairs has been reduced from 700 to 70.

You have already considered, earlier in the chapter, why birds of prey are naturally not as numerous as some other birds. Now it seems that they are being further reduced.

The use of chemicals in agriculture to control pests on crops is certainly one of the main causes of the decrease in the numbers of many of the birds of prey. These chemicals may poison the birds themselves or they may cause them to lay eggs which do not hatch.

Q10   In what position does a bird of prey come in a food chain or a pyramid of numbers?

Q11   Can you work out how a chemical sprayed on a crop can eventually be the cause of the death of a bird of prey?

Q12   What other reasons can you give for birds like the golden eagle and the peregrine falcon decreasing in numbers during the last hundred years?

Figure 228
Straw stubble being burned.
Photograph, Godfrey Bell.

Figure 229
A farm field where a mature
hedgerow is being removed.
Photograph, Godrey Bell.

Figures 228 and *229* show how man may sometimes destroy a whole habitat.

Q13 Do you think that such habitats should be protected? Try to give reasons for your answer.

Q14 Can you think of any reason why man should continue to be allowed to burn straw stubble and remove hedgerows?

You have now been given a few examples in which man has been responsible for destroying the balance in nature. Sometimes it has been accidental, sometimes deliberate, although he may not have realized how serious the results of his actions were going to be until it was too late.

Q15 What other examples do you know of how man, by his activities, has upset the balance in nature? You may know of examples in your own area as well as some which are of worldwide concern.

Q16 For all the examples you now have, can you decide whether the balance was upset because of deliberate acts of man or whether this was accidental?

If man has played such a part in upsetting the balance in nature, what is he doing to restore it or to protect it for the future?

11.3 Conserving the balance in nature

Nature conservation may be a familiar term but the meaning of it may not always be fully understood. It is often thought to mean just preservation and protection of animals.

But nature conservation means the management of *all* natural resources, which include water, soil, timber trees, and all other plants, and animals, so that where possible the balance is improved as well as kept.

Q1 How do you think man could have acted differently in the examples shown in *figures 220* (page 274). *221b* (page 275), *223* (page 277), *225* and *226* (page 278), *228* and *229*, so that the balance would not have been upset?

You may not have found it possible to make a satisfactory suggestion in all cases although sometimes the solutions to these problems are fairly straightforward. For instance, it is possible for a chemical used to destroy pests to be banned, or for its use to be controlled, if it is known to cause serious harm to other animals. But even this will not be a popular decision with everyone.

Q2 Which of the problems given in the examples would the banning or controlled use of harmful chemicals help to solve?

Q3 Do you know of any chemicals whose sale and use are controlled? See what you can find out about this.

At other times, it is not easy to know what decision should be made. For example, soap never caused a pollution problem, because it can be broken down by the action of bacteria in the sewage beds. On the other hand, detergents have better cleansing properties than soap. However, the most commonly used detergents in this country are *not* broken down by bacteria during sewage treatment.

Therefore, when the sewage effluent is released into a river, it may cause foaming and may have a harmful effect on the plants and animals living in the river.

Q4 Obviously we want to clean up the rivers but is it sensible to ban the use of detergents altogether?

We have taken only a very few examples of how man has been, and still is, responsible for upsetting a balance in nature. There are many other examples and you may have an opportunity to find out something about them.

It may not always be easy to decide what should be done to restore the balance or to prevent it from being disturbed again although, in some cases, better planning and more knowledge would have helped. Sometimes there are alternative ways for man to achieve his object without upsetting the balance. There is an example of such a way in Chapter 9. In fact the measure described sees that a proper balance is kept. Find out what this is.

Man differs from all other animals in that he not only has a memory but he is also able to think and reason. That is to say, he is able to plan what he is going to do in the light of his experience, before he actually does it. Before we spray our roadside plants, or pump sewage into our rivers, or take birds' eggs, we should pause to think what the results of our actions are likely to be. In other words, we must develop and use our ability to reason because, in the long run, we are the people who will bear responsibility for the destruction or the preservation of the places in which we live. The countryside, as well as being a source of food, is also there for our enjoyment, and we should be aware of how we can be an influence for good (or bad) in its preservation.

'As dead as the dodo'

Today the expression 'as dead as the dodo' is a reminder that there have been animal species of which there are now no living specimens. The dodo was a very odd bird which might be described as an oversized dove and it was, perhaps, doomed for extinction without any help from man. But this is not always the case with animals which have become extinct.

Penguins are thought of today as birds of the southern hemisphere. But there was once a 'northern penguin', better known as the great auk, which, like the penguin, could not fly and was awkward on land.

Although hunted by ancient man, the great auk once existed in millions but eventually it was only to be found in Newfoundland, Greenland, Iceland and the nearby islands, and the British Isles. It was 'discovered' by fishermen in the fifteenth century who killed great auks in large numbers for their meat. This tasted good and could be salted for keeping. They also collected the eggs, which were considered a delicacy. As a great auk laid only one egg each year, egg collecting was very disastrous indeed.

Figure 230
The great auk.

After about two hundred and fifty years of this egg collecting, the great auks survived only on the small rocky islands near Newfoundland, Greenland, and Iceland and on St Kilda, an island off the coast of Scotland. The last great auk on St Kilda was killed in 1821 and then these birds only remained on a small group of rocky islands known as Auk Rocks near Iceland. In spite of this their eggs were still collected until in 1830 a volcano erupted under the sea and this last home of the great auks disappeared.

When it seemed that it was too late, museum directors realized that there were very few great auks among their collections of stuffed birds.

Then it was discovered that a few great auks had escaped the eruption and were living on an even smaller island nearby. You might have thought that the museum directors and other scientists would have seen to it that the last of the great auks were protected in order to save the species. But instead they were hunted more fiercely than ever and vast sums of money were offered for them and their eggs. Between 1830 and 1844 all the remaining birds were hunted until in June 1844 the last two great auks were killed.

This meant that although the museums have stuffed great auks on show in their cases the species is gone for ever.

Figure 231
Ne-ne or Hawaiian geese.
Photograph, E. E. Jackson, The Wildfowl Trust.

Figure 232
The giant panda.

Less than a hundred years later it seemed that history would be repeated in Hawaii. The ne-ne or Hawaiian goose had become very rare on the island for much the same reasons as caused the great auks gradually to disappear. When, in 1911, less than one hundred wild ne-ne geese still existed it was finally decided that this bird must be protected. Even so their numbers still decreased until, in 1949, less than fifty of these handsome geese were left.

It was at about this time that the Wildfowl Trust at Slimbridge in Gloucestershire obtained three of these rare birds and went to great trouble to try and breed them. Hens and bantams were used to rear them and eventually, in 1955, a flock of these birds had been built up. The Wildfowl Trust was then able to send ne-ne geese back to Hawaii where they were released and have been established again.

This may have been the first time a rare bird has been saved from extinction by rearing some individuals in captivity and returning them to the wild. Recently, birds of prey, which are becoming rare, as we have seen, have also been bred in captivity and released in the wild. Eagle owls, bred in the Norfolk Wild Life Park, have been released in Sweden where they were becoming scarce.

Mammals which are in danger of extinction usually get more publicity than other animals. The part played by zoos and wild life parks all over the world in conserving these species is very important indeed.

Because of the attention given to the larger animals it is sometimes difficult to remember that conservation is concerned with all animals and plants. The beautiful large copper butterfly was known in eastern England until the middle of the nineteenth century. About this time many of the fens were being drained, which meant that some plant species had their habitat destroyed. One of the plants affected was the great water dock, which was the food plant of the large copper butterfly. The activities of collectors and the scarcity of its food plant were responsible for this beautiful butterfly becoming extinct in England. However, it still survived in the Dutch fens.

Figure 233
The female of the large copper butterfly.
Photograph, George E. Hyde.

About eighty years after it had become extinct in England, the large copper butterfly was re-introduced to Woodwalton Fen in Huntingdonshire where, owing to careful management of the water level in the fen, the conditions were restored so that the great water dock grew there again. Now it is no longer deprived of its food plant, the large copper butterfly has been successfully re-established in this country.

Many people get great pleasure from wild flowers and some pick them, or even uproot them to plant them in their own gardens, only to be disappointed when they do not grow and look as they did when growing in the wild state. (Incidentally, most counties in this country have a bye-law which forbids the uprooting of wild plants.)

The lady's slipper orchid is one of the rarest plants in this country. It has a very striking flower, with a yellow lip contrasting with the maroon or claret colour of the petals and sepals. This plant, which grew naturally in Yorkshire and Durham, has been very rare for a number of years, largely because of collectors digging it up and planting it in their own gardens, where it usually died. At one time, plants which had been dug up were being sold in Settle market in Yorkshire.

Figure 234
Lady's slipper orchid, *Cypripedium calceolus*.

You have seen how living organisms are to be found in a particular habitat. If they are moved to a completely different one, it is not surprising that they do not survive. The habitat of the lady's slipper orchid is the limestone soil to be found in the Dales of Yorkshire and Durham.

Fortunately one or two lady's slipper orchids are still known to exist but very few people are ever likely to see this beautiful plant growing wild in this country as its whereabouts is now a carefully kept secret.

When a plant is rare this always makes it more interesting. There are people who like to collect specimens of rare plants although, fortunately, it is more usual today to keep a photographic record.

When it was first suggested that a part of Upper Teesdale should be flooded to form a reservoir to supply water to large industrial works, botanists and many other people tried to prevent it.

Figure 235
The Teesdale violet, *Viola rupestris*.

The Teesdale violet was one of the interesting plants to grow in the area and there were others which formed a unique community of plants which grew in this country just after the Ice Age.

Many enquiries were held but the area has been flooded. It is possible, however, that some of these species still survive nearby. Certainly, the result of flooding the area was realized before it was too late but it was decided that there was no alternative site for the reservoir.

Index